Modern Sawmill Techniques

Volume 3

Other volumes in the Sawmill Clinic Library:

Modern Sawmill Techniques Volume 1: Proceedings of the first Sawmill Clinic, Portland, Oregon, February 1973

Modern Sawmill Techniques Volume 2: Proceedings of the second Sawmill Clinic, New Orleans, Louisiana, November 1973

***Modern Sawmill Techniques Volume 4:** Proceedings of the fourth Sawmill Clinic, New Orleans, Louisiana, November 1974

Volumes in the Plywood Clinic Library:

Modern Plywood Techniques Volume 1: Proceedings of the first Plywood Clinic, Portland, Oregon, February 1974

***Modern Plywood Techniques Volume 2:** Proceedings of the second Plywood Clinic, New Orleans, Louisiana, November 1974

Other Miller Freeman Publications books for the forest products industry:

An Introduction to Paper Industry Instrumentation by John R. Lavigne

***Fiber Conservation & Utilization** edited by Paul D. Van Derveer and Ken E. Lowe

Plywood Manufacturing Practices by Richard F. Baldwin

Timber Cutting Practices (rev. ed.) by Steve Conway

***Transport & Handling in the Pulp & Paper Industry** edited by Leonard Haas and John E. Kalish

*Scheduled for Spring 1975 publication

Modern Sawmill Techniques

Volume 3: Proceedings of
the third Sawmill Clinic,
Portland, Oregon, February 1974

Edited by Vernon S. White

Volume 3 of the Sawmill Clinic Library
A FOREST INDUSTRIES book

Library of Congress Catalog Card Number 73-88045
International Standard Book Number 0-87930-026-4

Table of Contents

	Foreword by the editor	7
	Authors & chapter summaries	11
1	Municipal steam/power from wood/other waste By S. LaMont Matthews	41
2	Western potential for yield/grade improvement By H. C. "Carl" Mason	52
3	Drying hogged fuel for powerhouse operation By Harry J. Hardy	65
4	Briquetting wood waste by the Hausmann method By Fred Hausmann	72
5	Bark breakdown for profitable products By Frank S. Trocino	91
6	Computerized planning for maximum profit By Dr. Eugene L. Bryan & John R. Lang	103
7	Orienting wood fibers for added strength By Ron Dorsett	119
8	Reliability of computers and scanners in the mill By Dan Pichulo & Lowell H. Hulbirt	128
9	Setting a precision carriage by computer By John D. Steadman	149
10	Water gun conversion for headrig carriages By Jim Bigelow	168

Table of Contents

11 Engineering for increased production — 171
By Donald E. Hand & Edward E. Washburn, Jr.

12 Measuring efficiency at mill stations — 190
By Wayne Bennett

13 Rube Goldberg gimmicks smooth mill flow — 211
By Melvin R. Paulus, B. T. "Burt" Randall, Larry L. Webb, & Richard L. DeArmond

14 Careful lumber handling cuts degrade losses — 235
By Dean W. Huber

15 Noise control: the compliance options — 250
By Frank N. Berglund

16 Noise reduction techniques for the sawmill industry — 257
By Dr. John S. Stewart

17 Narrow lumber planer designed for smaller logs — 276
By Gilbert W. Anderson & R. Brian Parsons

18 Save wood/raise grade with abrasive planing — 290
By Sherman E. Kirchmeier

19 CRT drying for reduction of degrade — 300
By Dr. Dallas S. Dedrick

20 Ultrasonics to detect wood's defects — 309
By Erwin H. Bulgrin

Exhibitors, 1974 Sawmill & Plywood Clinic, Portland, Oregon — 319

Index — 322

Foreword by the editor

A dramatic shift in the direction of the lumber industry's rapid technical change took place in the one short year between the Portland Clinics of February 1973 and February 1974.

That formerly rare word *shortage* became common conversational coin. With its appearance, lumbermen recognized that they would have to furnish–from their own waste–much of their future process steam, drying gas, and, yes, even electrical power. In being blessed with the capability to do that, the forest products industry must be nearly alone.

So the shortages, shutoffs, and spiraling prices of natural gas, fuel oil, gasoline, and hydroelectric power put urgency into an area which had stood far down the line in priority in the minds of most mill men. And the real value of wood waste as fuel became apparent as the spigot on natural gas sputtered off for the plywood plants.

So the practical question of conversion from wood fiber to Btu came to the fore. Hog? Hammermill? Steam boiler–how large? Dryer gas? Sander dust conversion? Air quality compliance? Fuel drying? How? At what cost? How must specific mixes of sawdust, shavings, and bark be handled to extract that precious heat? Shall we run the dryer, the kiln, the machinery; shall we produce electricity?

The 1974 Portland Sawmill Clinic sought to focus on these very real questions, but to do so without neglecting the technical revolution well underway in measuring logs accurately, in computerizing the decisions on their best use, and in cutting the products with the least waste.

And since 1973, the scanner-computer team had moved from what many considered a fantasy into reality. Mill men who had this team working for them were able to tell other mill men how they really worked.

The 1974 Clinic was expanded with the addition of a day and a half for plywood—a cooperative venture between Miller Freeman Publications and C. C. Crow Publications. Its president, Bob Smith, ably orchestrated Crow's effort, while Dave Pease, editor of CROW'S FOREST PRODUCTS DIGEST, built and presided over the plywood program.

Many registrants heard both programs. The longer Sawmill program dealt with many problems common to both plywood and lumber mills—sorting, log handling, waste conversion.

Addition of the Rube Goldberg panel to the 1974 Sawmill Clinic program proved an interesting highlight. In the emphasis placed on cutting stations, my feeling was that the handling of logs, cants, and lumber between stations gets too little attention from top management and, often, from the manufacturers who furnish the sophisticated cutting equipment. But there is a core of mill superintendents, millwrights, sawyers, edgermen, and the like who *must* get that job done exceedingly well if the mill is to operate at its optimum. The saw must be neither flooded nor starved. The wood must be presented to it in the right position; it must be transferred, turned, separated—above all, *moved*! So the world of roll cases, transfer tables, crowder rolls, surge decks, flop gates, and board dealers becomes as important as the saws themselves. Without these, the saws cannot function properly.

The growth of the machinery show adjoining the Clinic auditorium reflected the interest generated at the 1973 Portland Clinic. Plans for the first Clinic included a couple dozen booths, and that Clinic opened with 48—all the hall would take. In 1974 some 190 exhibit spaces were taken by 140 manufacturers serving the forest products industry—quadruple that first year's toe-in-the-water effort. And locating the exhibits in a separate hall eliminated the noise problem encountered at the 1973 Clinic.

Clinic attendance also has grown dramatically—from 835 registrants the first year to 1,300 the second. This growth reflects industry desire for information and for a marketplace to compare ideas. Such an opportunity has existed for more than 60 years in the logging side of the industry in the form of the Pacific Logging Congress and its regional conference spinoffs, many of them past 30 in age.

But the lumber industry did not have such an exchange. Somehow the political and economic differences between bigs and smalls, greens and drys, dimension and cutting, water and rail had reflected themselves in the associations, logical sponsors for such a gathering.

But certainly a publishing company serving an industry accepts the obligation to furnish information to that industry—and to do so in whatever form best suits the industry's needs.

Getting an exchange moving within the industry is the key.

So thanks are due the 15 men who addressed the plywood sessions and the 27 who contributed their thoughts to the Sawmill Clinic. Obviously, they were the heart of the operation. This book records the talks given at the 1974 Portland Sawmill Clinic and the questions elicited by those comments. Considerable illustration drawn from the material used at the Clinic has been included. Volume 1 in the new Plywood Clinic Library reproduces the plywood sessions.

We hope Volume 3 of the Sawmill Clinic Library will serve the reader as a guide to where the industry may be going. But it is the readers themselves who will determine where their industry actually goes. We feel certain this book can help.

Vernon S. White
Editor
July 1974
Portland, Oregon

The Clinic luncheons offer registrants a change of pace—while continuing the information exchange.

Clinic Authors & Chapter Summaries

CHAPTER 1 MUNICIPAL STEAM/POWER FROM WOOD/OTHER WASTE

By S. LaMont Matthews

S. LaMONT MATTHEWS was born in Nebraska and graduated with a B.S. in mechanical engineering from Oregon State University in 1956. In that year he started work as a project manager for Cornell, Howland, Hayes & Merryfield and Hill—probably the best-known engineering firm in the Northwest. He is now their director of mechanical engineering and, as such, has supervised the design and construction of a broad range of industrial and governmental projects. These include fuel handling and storage, boiler control, sanderdust firing systems, as well as numerous municipal and building projects. He holds engineering licenses in five states, and is a member of the American Society of Mechanical Engineers, American Society of Heating, Refrigeration and Air Conditioning Engineers, Air Pollution Control Association, Consulting Engineers Council of Oregon, and the Forest Products Research Society.

CHAPTER SUMMARY: Municipal electric power plants can be operated profitably even if no solid wood waste is available. A refuse-fired plant for the Eugene-Springfield, Oregon, area, for instance, would produce 35 megawatts and break even to start. Therefore plants solely dependent on municipal waste at first should be built with provision for expansion to include solid wood waste.

Much of the wood waste which mills were willing to commit earlier has now become too valuable for mills to be willing to make long-term commitments. Mill wastes have become by-products or a direct source of energy for mill steam or electrical generation. Nevertheless, there is strong feeling in the industry that some wastewood will be available to municipal plants in areas of heavy concentration of sawmills and plywood mills.

Mills may find that selling wastewood to the regional energy plant and buying back power and/or process steam may be more economic than building their own boilers or generating facilities.

The study for central Lane County, Oregon (Eugene-Springfield), envisioned a 130-acre plant with four 700-foot-diameter hogged fuel piles with storage of 260,000 units. Three wood waste-fired boilers would each produce 460,000 pounds of steam per hour. Additionally, two refuse-fired boilers would produce 108,000 pounds per hour each. This would supply a 680,000-pound-per-hour process steam load and generate 115 megawatts of electricity. Plant cost would be $55 million.

Central Lane County could start with 700 tons of solid waste daily and increase to 1,000 with population growth. About 83 percent is combustible. After removal of metal and glass, it is mostly shredded paper and can supplement hogged fuel boilers.

Even without mill wood waste, energy systems will be more economic than landfills. After a first break-even year, revenue from power sales would even be available to buy mill wood waste. The plant might have to charge $0.21 per unit to accept wood waste the first year, but should be able to pay $2.50 per ton for it within 10 years.

CHAPTER 2 WESTERN POTENTIAL FOR YIELD/GRADE IMPROVEMENT

By H. C. "Carl" Mason

H. C. "CARL" MASON, president and technical director of H. C. Mason & Associates, was raised in the Douglas fir logging and lumber community of Mapleton, Oregon. In 1943 he entered the University of Oregon on an athletic scholarship, majoring in physics. Two years later, he joined the U.S. Navy, where he continued his education in electronic engineering. He returned to the lumber manufacturing industry as an engineer and contractor of sawmills and other wood conversion plants. In 1953 he joined Columbia Lumber Co. in Juneau, Alaska, as a consultant, and the following year acquired Fryer Corp. of Portland, Oregon. He then began to expand into the design, engineering, and construction of sawmills, veneer plants, and plywood plants. The growing importance and demand for economic feasibility work led him to establish H. C. Mason & Associates in 1963. Through his innovative work as a consultant, Mason is recognized as a pioneer in the application of mathematical solutions to log sawing problems. His affiliations include memberships in the Institute of Electrical and Electronic Engineers, the Institute of Management Consultants, Association of Management Consultants, the Plywood Research Foundation, and the Southern Forest Products Association. He is a member of the board of the Western Environmental Trade Association and a former member of the board of Rotary International. He is also a graduate of the Advanced Management Institute of the Harvard Graduate School of Business Administration.

CHAPTER SUMMARY: The Pacific Northwest is one of the few areas of the world capable of sustaining production of high-grade softwood clears and timbers of identifiable species in reliable grades and continuing quantities.

Mason states that many mill operators in the region do not understand how to improve the relationship between yield and grade. Yield is expressed as the overrun, lumber tally, of mill input in Scribner log scale, expressed in board feet. This wood recovery varies with product (that is, 72 cubic feet of wood produce 1,000 board feet of scant-sawn 2x4, while it takes 100 to 120 cubic feet to get 1,000 board feet of full-sawn selects).

Yield results from: (1) product line requirements; (2) oversawing to provide for variations in sawing; (3) kerf, planing, and shrinkage allowances; and (4) breakdown method.

Product line should be the best mix of products for a given grade and species to yield maximum monetary return.

Sawing variation depends on machine accuracy. There is a great difference in yield according to sawing method. The sawmill industry understands how to cut dimension lumber accurately and scantly, but has not applied this skill to clears and uppers. Differences in yield from various cutting methods are shown.

Live sawing will show the greatest recovery when cutting dimension, except for avoiding spike knots. A small log which can produce only one 2x4 is unaffected by any fiber-saving factor. The larger the log, the greater the percentage by which fiber-saving factors affect recovery. These variables will give a predictable result when applied to any log class.

Grade cutting occurs only in western old-growth Douglas fir, hemlock, pine, true firs, redwood, spruce, larch, and cedar. Optimum recovery of yield often is incompatible with best grade recovery. Best possible grade recovery on bandsaws has been deterred by (1) capital investment, (2) headrig kerf, and (3) headrig carriage and bandsaw inaccuracy caused by impact shocks and equipment abuse.

Downstream equipment improvements are minimizing the number of headrig lines. Most useful are precision solid linebar and twin band resaws. Computer-controlled ball screw carriages are repeatedly accurate. Digital readouts permit sawyers to think only in nominal sizes.

Raw material is 50 to 80 percent of industry production cost. Processing costs are increasing more slowly than raw material costs. Wood has been badly treated in our economic system, but the author asserts it will assume its rightful place as the only renewable resource.

CHAPTER 3 DRYING HOGGED FUEL FOR POWERHOUSE OPERATION

By Harry J. Hardy

HARRY J. HARDY was born in Virden, Manitoba, Canada, and received his education in mechanical engineering from the International Correspondence School in 1949. He served as a fighter pilot in the Royal Canadian Air Force during World War II, completing 97 missions. From the end of World War II until 1960, he was employed by Flavelle Cedar Company, at which time he moved to MacMillan Bloedel, Ltd. There he worked as project engineer and senior design engineer until 1969, and is now engineering supervisor of the Canadian White Pine Division. A long-standing specialty has been the design of sawmills and plywood plants. An enthusiastic aviculturist, he reports 32 species of pheasants, waterfowl, and doves under his care.

CHAPTER SUMMARY: MacMillan, Bloedel, Ltd. installed a hogged fuel preparation system at its huge Canadian White Pine Division mill at Vancouver, British Columbia. The purpose was to minimize fly ash emissions from the powerhouse stack.

The objective of the system was to prevent formation of fly ash, rather than deal with it later—as two other systems at the mill attempt to do. The system is expected to meet Vancouver's municipal emission standard of 0.15 grains per standard cubic foot of flue gas.

Hogged bark is 70 percent water. A $250,000 bark press had reduced water content to 60 percent. The new $1.2 million Energex dryer will drop it to 44 percent. So removing 47 percent of the water cost $250,000, but the next 30 percent cost $1.2 million. The former was squeezable external water, the latter must be evaporated by heat.

To reduce water content fuel will (1) get secondary hogging to ¾-inch minus particles and (2) be dried from 60 percent to 44 percent moisture in a rotary dryer. This should produce more complete combustion and less fly ash. Reduced fuel water content will also improve boiler operation.

Hogged fuel moves from the mill through the bark press, up a conveyor, across the powerhouse, and into distributing conveyors over the furnaces. Excess fuel returns to the hogged fuel pit.

There is a metal detector between the hogs. A screw at the bottom of the bin meters 63 tons per hour into the powerhouse conveyor.

To dry, fuel enters a 9½-foot-diameter by 60-foot-long rotating dryer at 1,200 to 1,600 degrees Fahrenheit. Discharge temperature is 250 degrees Fahrenheit. The blower pulls 68,000 cubic feet of drying gas through per minute. About 31,000 pounds of water are extracted per hour and discharged to the atmosphere.

The hogged fuel produces about 2½ tons of salt daily. Stack temperature must be kept above salt dew point to prevent condensation. Discharge temperature controls the Energex heat output.

Fines are conveyed via an air-lock screw to storage from the discharge end of the dryer and from cyclones. These fines make fuel for the Energex, as does sander dust from the planer mill cyclones. The bottom screws in the storage bin charge the metering bins, which supply two 27½-Btu-per-hour Energex burners.

The Energex cyclonic burner is 36 inches in diameter by 60 inches in length and burns dry air and fines at 2,700 degrees Fahrenheit. Ten pounds of sand is also produced daily, most of which is recovered from the rotating screen at the discharge end of the dryer. It is blown back through the Energex, melted, and recovered as slag.

CHAPTER 4 BRIQUETTING WOOD WASTE BY THE HAUSMANN METHOD

By Fred Hausmann

FRED HAUSMANN is a native of Switzerland and president of Fred Hausmann, Ltd., formerly Glomera Holding Ltd., of Basel, Switzerland. For the past 19 years, his overwhelming concern with what he saw as the potential productivity of industrial wastes caused him to concentrate on the problem of waste utilization. In these wastes, he could see huge amounts of potential energy lying unproductive. This led him to invent briquetting devices which convert a large variety of residue materials—such as bark, sawdust, pulp fiber, almond hulls, peanut shells, aluminum and copper swarf, and so forth–into useful sources of fuel. Marketing these devices has taken him around the globe. A physical fitness enthusiast, he skis cross country, swims, water skis, climbs mountains, plays tennis, and combines his love of the outdoors with photography.

CHAPTER SUMMARY: Wood waste briquetting began during the 1920s in the U.S., and even earlier in Europe. Much financial success–including that of John D. Rockefeller with Standard Oil–is based on recovery and use of waste.

Today's high-pressure briquettors use pressures of 17,100 pounds per square inch. In today's machines, high-quality bearing parts and a thorough lubricating system are imperative. The 152-ton pressing power of the two rams requires precision in machine construction.

Many European briquetting installations work unattended. A number of forest products firms store wood waste as briquettes in summer to supply winter fuel needs. Savings result from supplying the firm's own heat and eliminating cost of waste removal.

The equipment life of a briquettor is estimated at 25 years, and installation cost is often recoverable in three years.

A Hausmann (or Glomera) briquettor uses a feeding device which partly compacts the material and prevents its backward expansion during the forward stroke of the ram. This improves the machine's capacity.

Briquetting is possible only when the wood's elasticity is broken. There is no agreement whether this is due to high pressure, to the heat formed, or to both. The heart of the press is the die holder. The cylindrical split die has four longitudinal slots. External hydraulic pressure changes the cylinder to a slightly conic shape–increasing log friction and counterpressure. The die holder has a water-supplied cooling coil.

Three types of briquetting installations are described and illustrated. In the simplest, fine dry material is fed directly from a silo through a vibrating chute, from a cyclone, or by a conveying system to the briquetting device. Maximum raw material moisture content is 15 percent.

Coarse dry material must be ground before briquetting.

Coarse wet material should be hogged, fed into a silo, and then to a dryer. Thermic drying is justified only for material with up to 55 percent moisture content. Above that, it should be pre-dried mechanically. Ring roller presses are most suited for pre-drying.

In conclusion Hausmann notes that the 21 million tons (dry weight) of unused bark and other wood wastes generated annually in Alaska, Washington, Oregon, and California could be converted into 11 million kilowatts of electric energy.

CHAPTER 5 BARK BREAKDOWN FOR PROFITABLE PRODUCTS

By Frank S. Trocino

FRANK S. TROCINO, technical director of Bohemia, Inc. of Eugene, Oregon, graduated in chemical engineering from the University of Iowa in his native state. He was plant superintendent of Grain Processing Corp. in Iowa, manufacturers of distilled alcohol, soy bean oil, and animal vitamins. Later, he served as process engineer for Borden Chemical, where he did research in adhesives as well as being responsible for the design, construction, and initiation of chemical plants. His work with Bohemia resulted in the first effective system for extracting chemicals from Douglas fir bark, now in operation at Bohemia's plant in Coburg, Oregon. He holds numerous patents and is a member of both the American Institute of Chemical Engineers and the American Chemical Society. Relative to his favorite pastime, he is a former Commodore of the Eugene Yacht Club.

CHAPTER SUMMARY: Douglas fir bark is composed of cork, bast fibers, and amorphous powder, which are all well defined in the bark. Douglas fir cork is more laminar in cell structure than Mediterranean oak cork. There is also more wax in Douglas fir than in most barks.

A solvent developed by Bohemia, Inc. extracts the wax, which is mostly in the cork—transforming the cork from laminar to cubical in cell structure, and introducing air pockets which "popcorn" the cork. Solvent is distilled from the wax and reused.

Similar waxes are used for polishes, carbon paper, and cosmetics, commanding $1.75 to $2.75 per pound.

Spent bark can be ground into a plywood extender, replacing corn cobs.

Cork can be sold granulated or in sheets for bulletin boards, paneling, and gaskets. Sheet cork brings $0.50 per pound.

Bast fibers give impact resistance to plastics and, at $0.45 per pound, can replace such imported competitors as sisal.

Amorphous powder is a phenol substitute for plywood adhesives.

Oregon and Washington generate about 5 million tons of bark yearly—compared with 14 million for the entire United States.

The bark plant Bohemia is building at Coburg, Oregon, breaks chunks of bark to 3 inches in diameter. Sawdust, limbs, and other wastes fire the plant's boiler. The bark is ground and then sent to a dryer, where it is dried to between 5 and 10 percent moisture content.

Wax content of Douglas fir bark is high. The wax is usable as an internal lubricant in plastics, competing with wax imported from Brazil. Other plastics require synthetic waxes which come from oil—now in short supply.

Douglas fir wax is hard and has been used in lipsticks. Bohemia Inc. is trying to upgrade it physically for use in carbon paper and plastics. Wax extracted from Douglas fir bark will provide the first domestic source of vegetable wax in the United States.

During extraction, only 0.1 percent of the solvent is lost—the rest being recycled.

Plant equipment at Coburg is similar to that of the oil-bearing seed industry and can be bought "off the shelf."

Bohemia's plywood adhesive extender has been approved by the American Plywood Association. Averaged in, it reduces glue cost. However, wax must first be removed from the bark.

The amorphous powder contains phenol, which is usable as a plywood glue—not just as an extender. APA approval is being sought. About 500 billion pounds of phenol-formaldehyde resin are used in the plywood industry yearly.

Processing will be adapted as raw material changes occur.

CHAPTER 6 COMPUTERIZED PLANNING FOR MAXIMUM PROFIT

By Dr. Eugene L. Bryan & John R. Lang

DR. EUGENE L. BRYAN did undergraduate work in forestry at the University of Idaho and graduated from the University of California at Berkeley with masters degrees in wood technology and engineering. He received a Ph.D. from the University of Michigan and world-wide recognition for his pioneering research in use of laser beams and high-energy water jets for wood cutting and machining. After a period as research manager for The Black Clawson Co., Everett, Washington, where he established the Wood Processing Laboratory, he joined the faculty of Humboldt State College in Arcata, California, as associate professor of forestry. In 1967, he left teaching for full-time consulting activities, with emphasis on computer applications for management decisions in the wood products industry. He is now president of Decision Dynamics Inc., in Portland, Oregon.

JOHN R. LANG is operations research analyst and manager of planning and administration with Evans Products Co.'s U.S. Wood Products Division. He holds B.S. degrees in physics and business administration from Oregon State University and completed business administration courses on the masters level at Arizona State University. He worked as a computer programmer and systems analyst for General Electric in Richland, Washington, and as a systems engineer and marketing representative at IBM in Phoenix. He is secretary-treasurer of the Northwest chapter of The Institute of Management Sciences, a member of Planning Executives Institute and the Forest Products Research Society. In his spare time, he manages his own commercial purebred cattle operation, featuring Italian and French breeds new to the United States.

CHAPTER SUMMARY: A sawmill manager has three concerns: raw material, production facilities, and market, Bryan points out. He needs to coordinate all variables in each category to obtain the best return on investment. However, the variables are so numerous that best performance cannot be obtained without computers.

A mill manager is affected by his internal plant environment, by the state of the industry, and by external conditions which he cannot influence but to which he must constantly adjust. Timely information and action are required. Computers can help keep the operation in context.

One computer-based aid is linear programming which can optimize allocation of resources for maximum profit.

Albert Einstein could juggle only six variables at once in his mind. Most men can handle fewer. Computers can integrate many more. Models can be created to study the effect of all variables. Computer models are very useful in making management decisions.

Lang relates how Evans Products Co.'s Canadian division increased profits of a dimension sawmill without buying an excessive amount of new equipment. This was accomplished with the help of a three-man task force and Decision Dynamics, Inc.

All key variables and decision points were built into a computer model. Points found critical to sawmill profits were: log input and lumber output from the chipping headrig; diameter determination and correct setting while maintaining 2,000-log-per-shift production. A log scanner with a minicomputer was installed to handle size differences not visible to the human eye. A plus is that the minicomputer also functions as a process control device.

The computer records pieces per shift more accurately than the crew. It records diameter and length, permitting accurate measurement of lumber recovery. Operator override of computer decisions is less than 10 percent.

Woods bucking policy was evaluated. Price premiums for various lumber lengths were fed through the computer to establish optimum bucking policy.

Bandmill headrig breakdown patterns and sorting divisions at the green chain were similarly analyzed.

Such decisions could be modelled by a computer program. Evans Products Co. is now studying whether specific logs are most valuable as lumber, veneer, or plywood—to achieve improved log allocation.

CHAPTER 7 ORIENTING WOOD FIBERS FOR ADDED STRENGTH

By Ron Dorsett

RONALD DORSETT, president of MK-RDA, Inc., acquired a degree in mechanical engineering from the Royal Canadian Naval Academy in 1946 and completed graduate courses in forestry and wood science and technology at the Forest Products Laboratory and the University of British Columbia. He started his professional career as a laboratory technician at the Forest Products Laboratory of the Canadian Department of Research at Vancouver, B.C. From 1949 to 1951 he was an inspector of ships and general engineering for the Canadian Department of National Defense. In 1957 he formed Ron Dorsett Associates, Inc., an independent consulting firm operating out of Portland, Oregon, and Phoenix, Arizona. In 1972, this firm embarked upon a joint venture, of which he was general manager, with Morrison-Knudsen Company, Inc. The two firms provided a complete range of services for the forest products industry, and Dorsett became president when they merged to form MK-RDA, Inc. in 1973. He is a member of the American Society of Mechanical Engineers, Society of Wood Science and Technology, Society of American Foresters, American Institute of Timber Construction, Technical Association of the Pulp and Paper Industry, American Society for Testing and Materials, and the Forest Products Research Society.

CHAPTER SUMMARY: The concept of wood conversion to solid products may be revolutionized by a process developed at Washington State University and commercialized for industry application by MK-RDA, Inc. of Portland, Oregon.

The Fiber Orientation Process reduces mill residues and other solid pieces to wood fiber. Then, by aligning the fibers, it returns to the product the engineered properties provided by nature.

MK-RDA and WSU Laboratory tests indicate engineered strengths equal to—and in some cases greater than—those of common lumber and plywood. Nonmerchantable tree species have been reduced to fiber and have produced structurally sound products. This is a dry electrical process using sufficient moisture to act as a conductor.

Fibers are polarized during free fall through an electrical field. They fall onto the mat in full alignment.

Three factors are critical: (1) slenderness ratio of fibers; (2) moisture content; and (3) electrical power. Fiber length should be not less than 10 times diameter. Moisture content varies with species and whether current is direct or alternating. Amperage is very low, but 100,000 volts are used.

The only area of a fiberboard plant affected by such an installation is the forming station. Capital costs to a new plant is no greater than for one lacking orientation capability.

With this strength, fiberboards will assume structural capability and may substitute in many applications for lumber and plywood. Formed wood panel products could substitute for many present plastic applications in aircraft and automotive industries. Orientation allows weight reduction in present uses.

CHAPTER 8 RELIABILITY OF COMPUTERS AND SCANNERS IN THE MILL

By Dan Pichulo & Lowell H. Hulbirt

DAN PICHULO heads marketing management at Atmospheric Sciences Inc. (ASI, Sunnyvale, California). He graduated from Tufts University in Massachusetts in 1954 with a B.S. degree in chemical engineering. After graduation, he joined Du Pont's Atomic Energy Division in Augusta, Georgia, as process construction engineer. He remained with Du Pont until 1969, ultimately becoming manager for systems marketing and sales at the company's Wilmington, Delaware, headquarters. In 1969, Pichulo moved to Los Altos, California, forming his own company in motivation management. He was president there until April 1972, when he joined ASI. Pichulo's responsibilities there include coordination of marketing and sales activities for the southern United States. Most recently, he has been active in development and marketing of electronic measurement and computerized process control systems now coming into use in sawmills. Pichulo is a strong advocate of computerized process control systems and the use of production information to assist management in making better and more comprehensive decisions.

LOWELL H. HULBIRT, a native of California, received his B.S. degree from the California Institute of Technology, Pasadena, and obtained an M.B.A. degree from Harvard University's school of Business Administration. As a management consultant, Hulbirt spent 15 years with Ernst and Ernst and 7 years with the Times Mirror Co. in its Management Services Division. As manager of management science, Hulbirt is now on the corporate staff of the Times Mirror, assigned to Publishers Paper Co., Oregon City, Oregon, a subsidiary of the Times Mirror Co. One of Hulbirt's interests in the management field is the realization of the full potential of the human resource in industry.

CHAPTER SUMMARY: Pichulo discusses operational experience with sawmill computer/scanner installations and the use of minicomputers for log accountability, management information, and saw control.

Publishers Paper Co. has installed its third quad saw computer/scanner. Three lumber tally systems were also installed. Log input/lumber output can be measured accurately. Another system will determine optimum bucking patterns. One Chip-N-Saw mill with automated setting reported a 17 percent overrun increase after installation of the system. Sorting was automated at two locations. One of these automated systems measures and classifies logs at 400 feet per minute and kicks them into bins.

The reliability of these systems is greater than is commonly believed. The use of computer/scanner installations has been adopted quickly by some companies, and more slowly by the industry at large. The minicomputer makes possible integration of all the variables in a mill operation—a task which the human mind in unable to perform, and one which is becoming increasingly important to mill operators.

Hulbirt discusses the maintenance of computer/scanner systems at Publishers Paper Co. installations at Molalla, Oregon, and Burney, California. The principal source of operating and maintenance problems is the mill environment, rather than the the machinery itself. These problems include: dust in the air—sawdust, bark dust, yard dust; moisture; vibration—particularly when the sensor is set close to heavy machinery; and light from outside sources.

Sensors operate through light reflected from the surface being measured. Location and shielding of sensors are important to avoid inaccuracies caused by outside light. Chips or other material falling on the rail with the log may also cause a misreading.

Positioning a scanner incorrectly in relation to a post or beam or a holddown wheel dropping into the scanner's range may cause mismeasurement or miscounting. Sensors should be protected from welding flash. Lenses and bulbs must be kept clean of dust and moisture.

Dust buildup is one of the most common causes of computer downtime. However, with regular cleaning, this can be kept to a minimum, and the computer can be expected to operate as reliably as any piece of equipment in the mill.

For maintenance, the mill can either obtain a service contract, at about 10 percent of installation cost, or train its own crew. Publishers Paper Co. has trained their crew for troubleshooting, but service people are called in to repair electronics inside the computer. Uptime at Molalla has been 80 to 85 percent; and at Burney, 90 to 95 percent.

CHAPTER 9 SETTING A PRECISION CARRIAGE BY COMPUTER

By John D. Steadman

JOHN D. STEADMAN, originally from Ohio, graduated from Case Institute of Technology with a B.S. in electrical engineering and did graduate work in electronics at Marquette University. His career began with Joy Manufacturing Co., where he was involved in electrical engineering relating to mining, construction, and industrial machinery. After a period with Louis Allis Co. in engineering and marketing management, he moved to Filer & Stowell. There, he was promoted from chief engineer responsible for development of precision carriages and headrig chippers in Milwaukee to West Coast district sales manager in Portland–his present position. He is a member of the Institute of Electrical and Electronic Engineers and heads for the outdoors in his spare time to camp, fish, and canoe.

CHAPTER SUMMARY: As long as trees are of nonuniform sizes and characteristics and grade is a factor, there will be a need for a headrig bandmill system including a carriage, log turner, feed mechanism and skilled sawyer.

A typical carriage and setworks is accurate to about plus or minus 1/32 inch. Age and lack of maintenance cause deterioration from that level.

A typical sawyer reads a backstand dial to only 1/4 inch—perhaps 1/8 inch. He then calculates runs to allow a minimum surplus.

The precision ball screw carriage eliminates the greatest possible number of mechanical drives and linkages, and uses antifriction or long-wearing devices.

A closed-loop setworks is used. The position sensor knows where it must stop. It decelerates ahead and holds position against any external force. A one-thousandth of an inch knee movement will produce countertorque.

A computer should be used with the precision carriage and setworks. The computer can produce a number of sets for various cutting programs and can calculate allowance for pieces to be resawn downstream.

All functions are repeatedly accurate. The sawyer is relieved of mental arithmetic calculations and can concentrate on grade. All data are presented in digital form, with kerf, and so forth, automatically included.

Open-loop systems are accurate to 1/32 inch about 85 percent of the time, but accuracy may drop to 5/32 inch. The worst possible deviation for a closed-loop setworks would be 0.007 inch.

A precision carriage system may require double or triple the investment for a typical open-loop system, but improvement in recovery is substantial.

A Louisiana purchaser reported he had calculated a probable overrun increase of 4 percent with the system—indicating a one-year payoff. The actual increase, however, was 10 percent and payoff took only six months.

CHAPTER 10 WATER GUN CONVERSION FOR HEADRIG CARRIAGES

By Jim Bigelow

JIM BIGELOW, western U.S. dealer for Beloit-Passavant's sawmill machinery (including the slabbing chipper, barkers, and other chippers), Gemaco water shotguns, and Superior C&D Setworks for all machines, was born in Seattle, Washington. He completed his education there after spending part of his youth in Spokane, New York, and Connecticut. He served in the Pacific with the U.S. Navy during World War II, coming out a lieutenant commander. He was closely involved in the development of portable spars for the logging industry. Long a dealer for the LeTourneau log stacker, Bigelow was a pioneer in introducing dry log handling on the West Coast and is familiar with Scandinavian sawmill practices. A devoted yachtsman, he has also completed a Baja roadrace through lower California.

CHAPTER SUMMARY: With energy becoming more critical, it is necessary for the mill to capture what is now generated there, but wasted.

One method of energy conservation is use of a water shotgun to power the headrig carriage. Compared with a steam gun, a 10- to 12-inch water shotgun will save up to 10,000 pounds of steam per hour.

Conversion from steam to water can be completed over a very long weekend. A good opportunity is when the steam gun needs reboring or reringing. These functions are unnecessary with the water gun.

Carriage response with water is reportedly faster, resulting in one or two additional lines cut per minute.

Water is positive—hence easier for the sawyer to control than steam. Water cannot be compressed, so there is no delay. Water is recirculated through a 1,000- to 1,500-gallon tank. Electric pump power is used only when the carriage is moving. Water temperature is controlled either by a fan or a water-cooled heat exchanger.

The heart of the gun is a patented valve controlled by a hydraulic slave valve. Maintenance is eased if the valve is mounted outside the tank.

CHAPTER 11 ENGINEERING FOR INCREASED PRODUCTION

By Donald E. Hand & Edward E. Washburn

DONALD E. HAND was born in Los Angeles, California, and was graduated from the Carnegie Institute of Technology, Pittsburgh, Pennsylvania, with a B.S. degree in industrial management and a minor in electrical engineering. He worked at Cutler-Hammer Inc. as the Portland branch manager and electrical systems application sales engineer, later moving to Industrial Components, Inc., as vice president and applications engineer. He is now vice president and director of marketing for Dyna-Tech Corp., Portland, Oregon. He is a member of the Forest Products Research Society and the Institute of Electrical and Electronics Engineers.

EDWARD E. WASHBURN, JR. studied digital computer technology at the Massachusetts Institute of Technology, modern shop practices at the American Technological Institute, and mechanical engineering through the International Correspondence School. He is the third generation of lumbermen in his family. Through consulting activities conducted by his firm, E. E. Washburn & Associates, he has studied, designed, and built sawmills in Costa Rica, Canada, and the U.S.–including Alaska. A recent assignment, described in this volume, involved working with Del Conner Lumber Co. in Darby, Montana, as a project engineer. Washburn is now associated with Dyna-Tech Corp. in Portland, Oregon. He is a member of the Elks and Lions clubs, indulges in golfing, hunting, fishing and flying, and is a member of the National Pilots' Association.

CHAPTER SUMMARY: Hand points out that one resource not yet properly utilized in sawmills is people.

We need to determine conditions which will permit employees to work most efficiently, and then provide them. Equipment height, operator's field of vision, seat configuration, location of controls, and adjustability of seats for the differences in men's sizes are all important.

The location of controls should be governed by machine function and frequency of use. Controls for functions common to many machines should be placed in a similar position on each control panel so operators can make easier transition from machine to machine.

Temperature, humidity, light, noise, ventilation, and vibration are factors in the operator's productivity.

Machinery should be designed for ease of maintenance, particularly as it becomes more sophisticated and as maintenance people are required to work on a greater variety of equipment–brute steel to complicated electronics.

As the forest products industry gains experience with electronics, the reliability of these systems will improve.

Del Conner Lumber Co. is processing logs to a 2½-inch minimum diameter (and a 9-inch maximum diameter). It is, therefore, a very small log mill. As sort spreads were 1 to 2 inches, a scanner was needed. Washburn describes the plant layout designed to maximize production.

It was determined that running a single diameter class butt-to-butt would increase production 42 percent over setting each log. Therefore, a size sort designed by Dyna-Tech Corp. was installed. Five cutting programs were developed. Production capacity became 1,020 logs per hour–a system achievable only with automation.

Logs are not measured in the woods, but are bucked on the deck at 2½ inches. A Valon Kone debarker then barks at 197 feet per minute. Although bucking waste is not barked, it is chipped, leaving chip mix under the 1 percent allowance for bark.

Barked logs are then scanned and the computer decides which of five bins will receive them. Feed to the chipping headrig is automatic, but involves a 6- to 8-inch gap between logs on the headrig infeed conveyor.

CHAPTER 12 MEASURING EFFICIENCY AT MILL STATIONS

by Wayne Bennett

WAYNE BENNETT holds an M.S. degree in mechanical engineering from Oregon State University. After graduation, he joined The Nestle Co. in New York as assistant chief engineer of process design and development. He returned to Washington to join Weyerhaeuser as laboratory director of its Longview Technical Center, and is now manager of lumber research and development at Weyerhaeuser's corporate headquarters in Tacoma. His interests center on sawmill systems analysis and process improvement. He is a member of the Forest Products Research Society and is on the Advisory Board of Washington State University's College of Engineering. An ardent whitewater driftboat fisherman with a penchant for taking sports cars apart, he still finds time to undertake various youth activities.

CHAPTER SUMMARY: To get maximum profit from a sawmill, the overall strategy must be correct and each subsystem or cutting station must be operated optimally.

Decisions must not only be correct, but they must be properly executed. Good decision making requires information about raw material, machine capability, and product line.

One objective in the mill is to minimize the sum of (1) kerf, (2) deviation from accuracy, and (3) roughness. Kerf may have to be increased to control accuracy, but the sum of the three must be minimized. Minimum use of wood per saw line cut is the goal.

Efficiency is affected by (1) ability to position the wood before the saw, (2) ability to convey it straight during cutting, and (3) accuracy of the cutting tool's alignment. This is affected by vibration and by power fluctuations. Speed of sawing greatly affects accuracy.

Photographic checking of a station avoids the "halo effect" which a personal observer has on a machine operator. It also allows the manager to see if the wood being thrown away should have been utilized.

Measurement of deviation continues to be done with calipers, but it is stated in thousandths of an inch rather than in fractions.

Measurement is needed in positioning. To get the best opening face, measurement must be accurate to about one-tenth of an inch. Work is needed on positioning logs and cants to the cutting tool. We have the strategy but lack the hardware.

To test the ability of feed rolls to feed an edger accurately, Weyerhaeuser painted a straight line down some cants, taped a yardstick to the machine frame, mounted a high-speed camera, and ran the cants full speed, without saws. Lateral movement was as great as 1¼ inches.

Vibration must be controlled. Taut-wire machine alignment may be replaced by optical lasers, which can align under full load and speed.

Weyerhaeuser uses a precision test mandrel on saws. Saw flutter can be as great as ¼ inch. Research on correct saw tooth geometry is needed. Other needs are for guides and their alignment, for precision and repeatability in setworks, and for instruments with which to obtain accurate measurement. Instrumentation is needed for use in decision-making, to monitor cutting station performance, and for use in equipment maintenance.

CHAPTER 13 RUBE GOLDBERG GIMMICKS SMOOTH MILL FLOW

By Melvin R. Paulus, B. T. "Burt" Randall, Larry L. Webb, and Richard L. DeArmond

MELVIN R. PAULUS, manager of Lebanon Machine Co., Lebanon, Oregon, for the past six years, was born in Phillipsburg, Ohio. He completed a course in business education at Linn-Benton College in Albany, Oregon. Paulus started out as a foreman of a construction crew, later becoming superintendent of Midget Lumber. He then moved to Lebanon Lumber Co. as head millwright. Paulus is a member of the Forest Products Research Society and chairman of his local school board budget committee.

B. T. "BURT" RANDALL, a native Oregonian, studied mechanical engineering at the University of Oregon and then at the University of California. He started his career at the Oregon Pulp & Paper Co., and later moved to Long Beach, California, to become involved in the oil industry as superintendent with Olympic Refinery. He was an engineer with Consolidated Steel Co. and manager of South Umpqua Remanufacturing Co. before joining C & D Lumber Co., first as machinery manager and later as mill superintendent involved with increasing production and modifying machinery. He is credited with having developed the first pushbutton edger setworks. He is now with Superior Plywood Machinery Co. as their sawmill consultant.

LARRY L. WEBB studied at Oregon State University and has been in the mill business since a very early age. Starting at Brown Brothers in 1963, he was a green chain puller, a pond monkey, truck driver, Cat driver, pond sawyer, barkerman, resawyer, grader, edgerman, trim sawyer, head sawyer, millwright, and sawfitter. Webb later became foreman and is now vice president of that firm (in Grants Pass, Oregon) under the name of Webco Lumber, Inc.

RICHARD L. DeARMOND, vice president of Idaho Forest Industries, Inc., was born in Bend, Oregon, and took a degree in business administration from Oregon State University. His business history includes vice presidencies of the DeArmond Brothers Lumber Co. in Rogue River, Oregon, DeArmond Joyner Lumber Co. in Coeur d'Alene, Idaho, and presidency of DeArmond Lumber Co., Inc. He is a member of Western Wood Products Quality Standard Commission and his local Hoo Hoo.

CHAPTER SUMMARY: Rube Goldberg was a cartoonist of the twenties and thirties who portrayed complex, impractical efforts to produce results which were not worth achieving in the first place.

However, while we try to get the best results with the least effort in the mill, we too often back ourselves into mazes and complicated corners from which it is difficult to emerge in a straightforward way. Since we cannot start over, we develop what devices we can to deal with a given situation. Paulus gives several suggestions. One is to stimulate the crew to think and to submit ideas. A man whose idea is in operation is a more enthusiastic member of the team. (A Lebanon Lumber millwright developed a four-saw scragg with all saws on one arbor. It doubled production capacity within two months.)

Other ideas concern fast infeed, roll edgers, skate wheels to line up lumber, pineapple rolls, and variable lumber lines for the planer infeed and outfeed.

Randall discusses bucking with electric eyes. This is quick, cheap, and eliminates the repairs necessary with mechanical stops. The mill should be able to turn incoming logs. There are advantages to running small-end first. Run the log up a small incline to a turn table. Where it drops over is the center of gravity. Rotate there.

Pick and store edgings to be fed on downstream during pauses in flow from the headrig. A feeder that can be stopped at will and a moderate amount of chain are the only needs.

Size should be monitored at the trimmer to check whether planing allowance is correct.

Accurate counts of piece flow and machine revolutions are necesary to determine how to optimize equipment and manpower.

The quad band edger at Webco Lumber Co. is the subject of Webb's presentation. It has 16-gauge saws with 1/8-inch kerf, runs 180 feet per minute, and will accept cants to 16 inches thick. Changing saws takes 10 to 15 minutes.

As there is a slabber in front of the headrig, it is possible to use the quad edger as a small-log mill. Logs below 10 inches are slabbed on one side, fed to the quad, and then to the resaw.

Defective, hollow white fir—which would break up on the headrig, stopping production—is run directly through the quad edger. Unusable material goes to the chipper.

The quad permits 1-inch cuts between saws—instead of 3-inch, as on a conventional edger. It requires only one pass.

DeArmond relates the experience of Idaho Forest Industries in dealing with pollution problems. At first, they spent much on burners which were inadequate. Then, the firm turned to Norman Sowards of Environmental Products, Inc., which developed a sand-base burning bed. It has no grates. Most fuel fed into the top flashes out. The sand bed is fluidized at 2,000 to 2,200 degrees Fahrenheit. A cyclone removes particles and dumps them into an ash bin. Exit gas from the burner is cooled to below 800 degrees Fahrenheit to avoid burning out the cyclones.

The burner at the Atlas mill has been shut down and the stud mill refuse is put through the new system. It is prestored in a 60-unit bin and continuously metered. A package boiler will be added later.

CHAPTER 14 CAREFUL LUMBER HANDLING CUTS DEGRADE LOSSES

By Dean W. Huber

DEAN W. HUBER's contact with the world of sawmills began early. His father is head sawyer at a Washington mill, and a summer's employment as a sorter operator and green chain puller at Weyerhaeuser in Longview, Washington, sharpened his personal enthusiasm for the industry. Later, he pursued his interest in kiln drying at the University of Idaho, where he received his M.S. in wood utilization in 1972. He developed various drying schedules for ponderosa pine and Philippine mahogany for Boise Cascade, and evaluated the concept of prefreeze seasoning of California redwood for Simpson Timber Co. His interest in lumber drying led him to Moore Dry Kiln's Technical Services Department. He also worked with Automated Combustion in development and application of an industrial-sized wastewood-fired furnace. He is now head of his own firm, Huber Consulting, a wood technology consulting firm serving the forest products industries.

CHAPTER SUMMARY: The increasing value of wood as it passes through the different phases of mill manufacture should not be wasted by carelessness at the "back end." Much lumber that is now degraded en route to the boxcar could be shipped as high-grade lumber with improved practices in trimming, sorting, stacking, and drying.

Trimmers are frequently flooded with too great a piece count for the men and machines to handle properly. Air circulation is vital to even drying. Lumber of uneven thickness crushes kiln stickers, clogs air passage, and results in degrade or off-moisture shipment.

Sorting to length groups including only a few pieces of the longest length results in wasted kiln space, twisting, and end-split.

Stickers need to be kept in line vertically to prevent weight of top courses from bending or breaking lower stickers. Employees need careful instruction in degrade-prevention practices. Attitudes upstream in the mill—as reflected in the condition of the lumber—will affect attitudes downstream.

Kiln car bunks should be aligned with sticker-stacker placement, so that stickers line up with bottom vertical support. Correct kiln baffling forces air through the lumber; lack of baffling wastes the effort.

What savings are possible by intelligent care? Some mills have added 35 percent to kiln capacity with speedier drying. End-trimming necessary to upgrade lumber has been reduced 15 to 20 percent in some cases. One stud mill reported a 3 to 5 percent increase in grade with improved piling practices.

Much improvement is possible when using these techniques to plug the dikes on degrade.

CHAPTER 15 NOISE CONTROL: THE COMPLIANCE OPTIONS

By Frank N. Berglund

FRANK N. BERGLUND has been concerned with safety and its related areas for over 20 years. He completed a safety engineering course at the University of British Columbia, and combined his knowledge with responsibilities in purchasing and accounting at the Tygh Valley Lumber Co. in Bend, Oregon. He was assistant personnel manager and safety director at Brooks-Scanlon, Inc., Bend, Oregon, and later at Timber Products Co. in Medford, Oregon, personnel manager and safety director. He then joined Roseburg Lumber Co. as safety director and industrial compensation claims coordinator, and is now safety engineer. His expertise was recognized by his appointment as chairman of the Hearing Conservation and Noise Abatement Committee of the Wood Products Section of the National Safety Council, of which he is still a member. At present he is also a member of the Occupational Safety and Health Committee of the Western Wood Products Association and the Forest Products Safety Conference. A real outdoorsman, hunter, fisher, he has time to give first aid instruction and keep active in the Cub Scouts and Lions Club and serve on a local school board.

CHAPTER SUMMARY: Industry efforts to cooperate with government regulations on noise control began with enforcement of the Walsh-Healy Act. The Northwest Noise Committee was selected as a name for some 60 firms, some of which began research on engineering techniques to control noise. By the second year of their work, their findings were disseminated through trade associations.

Roseburg Lumber Co. found sound measuring devices which were temporarily accepted by government. A plant inspection there showed some clear violations, but also some high noise areas where workman exposure was of acceptably short duration. However, Roseburg Lumber Co. was not affected by the Walsh-Healy Act as it had no government contracts.

This, however, was only a prelude. The Occupational Safety and Health Act (OSHA) was then passed.

Management hopes to develop a means of testing sound levels that will be accepted in court as a defense against claims of hearing loss due to the mill environment. Data collected with present sound measuring devices are not adequate evidence.

The principal sound control methods now used include: (1) isolation of the machine and/or the operator from noise areas, (2) job rotation—which is rendered difficult by union contracts and wage rate differentials, and (3) ear protection.

Efforts to lower the acceptable decibel maximum to 85 are underway, but a current study may show this to be impractical with present technology. Western Wood Products Association plans to hire a consultant to study selected mills. The results will be presented to legislators.

CHAPTER 16 NOISE REDUCTION TECHNIQUES FOR THE SAWMILL INDUSTRY

By Dr. John S. Stewart

DR. JOHN S. STEWART, director of Noise Control Services, Inc., Greensboro, North Carolina, is a native of that state. He graduated from North Carolina State University with a Ph.D. in mechanical engineering, emphasizing sound and vibration research. He was a NASA trainee and research associate in acoustics at the same university and was an acoustical research engineer for Newman Machine Co. during the summer of 1971. From 1969 to 1972, he was a research associate at NCSU's Center for Acoustical Studies. An author of numerous articles and frequent contributor to professional society proceedings, he is now a member of the North Carolina State University faculty, Department of Mechanical and Aerospace Engineering. Stewart is a member of the Acoustical Society of America, the American Society of Mechanical Engineers, the North Carolina Acoustical Society, as well as several national honorary societies.

CHAPTER SUMMARY: To reduce machinery noise in the sawmill, efforts should be directed at the machine or its near field, since barriers and acoustical wall and ceiling treatment generally are not sufficiently effective.

Vibration isolation of the machine from the floor usually reduces only low-frequency noise components.

Well-located acoustical baffles do reduce noise a little. Enclosures should have sound-reflecting, as well as sound-absorptive, material, as several reflections of a sound wave are needed for effective absorption.

Rigid, sealed enclosures are less effective than acoustically lined enclosures that are isolated from machine vibration.

Noise-reducing tooling includes special cutterheads and saw blades. Idle cutterhead noise results from the periodic passage of knives past stationary surfaces. Cutting noise results from dissipation of energy stored in the piece being worked. Knife impact can be influenced by segmented cutterheads, helical knives, and other design parameters. However, noise-reducing cutterheads in sawmill planers cannot reduce noise to legally acceptable levels. Adding a compact partial enclosure can.

Circular saw noise comes from resonant blade vibration, air flow turbulence, or their interaction. Several techniques are practical in reducing noise to acceptable levels. Most effective are blade lamination and the laminated collar approach. Acoustical enclosures and lower saw speeds help.

CHAPTER 17 NARROW LUMBER PLANER DESIGNED FOR SMALLER LOGS

By Gilbert W. Anderson & R. Brian Parsons

GILBERT W. ANDERSON, a Seattle native, graduated in economics from the University of Washington in 1952. He is a veteran of both World War II and the Korean conflict. For 13 years he was with United Control Corporation, where he reached the rank of vice president and general manager. He was named president of Stetson-Ross Machine Co. Inc., Seattle, Washington, in 1970. He is a member of the Safety Standards Committee of the Western Machinery Manufacturers Association. One of his particular interests is development of foreign markets for wood processing machinery. He has travelled in the Far East, Scandinavia and Europe in support of foreign sales and licensing, and attended the 1973 Hanover Fair in Germany. He was one of the first Americans to get a look at industrial Russia through his participation in Lesdreumash 73.

R. BRIAN PARSONS, senior project engineer for Crown Zellerbach Canada, Ltd., and a native of British Columbia, started out as an engineer at Sorg Pulp and, later, as an engineer in wood handling and materials handling for H. A. Simons (International) Ltd., both in British Columbia. In 1955, he joined the staff of Crown Zellerbach Canada, specializing in design and construction of sawmills, plywood plants, woodmills, pulp mill wood handling systems, materials handling systems, barking plants, and chipping plants.

CHAPTER SUMMARY: Anderson describes the operation of a new narrow planer for small-log production lines, which has been developed by Stetson-Ross Machine Company of Seattle, Washington. Older planer-matchers have great capability to handle large material but rarely touch a piece larger than 2x12. With more small logs, a fast, easy-to-maintain narrow lumber planer requiring a lower capital investment was needed.

The resultant narrow lumber planer is not just a stud planer. Its three models have capability to cut 4x4, 4x8, or 4x12. The common frame permits changes in size capability.

Feed speeds vary from 300 to 1,200 feet per minute. Maintenance is easy. Cylinders and feed rolls are supported and driven from one side of the frame only. The operator's hands need never enter the cutting area when jointing.

Cantilevered feed rolls and cutterheads are on tapered spindles, making changes easy. Smaller size reduces cutterhead noise.

A Stetson-Ross planer designed in 1945 is run beside the new narrow lumber planer at Crown Zellerbach's Armstrong, British Columbia, mill. The former appears to labor at 450 revolutions per minute while the narrow planer almost seems to loaf at 800 revolutions per minute.

The chapter also includes the narrative of a movie on the high-speed lumber sorting system installed by Stetson-Ross for Biles-Coleman Lumber Co., at Omak, Washington. The system runs 100 lugs per minute, so the lumber line is split transversely into three lines for grading, then rejoined.

Parsons describes the Crown Zellerbach mill at Armstrong, in the interior of British Columbia, which was designed to produce 2x4 only. Some 2x3 and 1x4 are generated as falldown, however.

Mill output is about 315,000 board feet daily (double shift). Some 240,000 board feet of random-length 2x4 must be dressed daily. This requires a sustained planer speed of about 800 feet per minute.

No high-speed, narrow machine could be found, and the purchase of a 6x15 planer could not be justified. Stetson-Ross was approached and agreed to design and build such a machine. The only departure from traditional planer design was use of stress-relieved weldments instead of castings.

The planer has operated successfully with minimum debugging. The demand for this machine evidently had existed before it was developed.

CHAPTER 18 SAVE WOOD/RAISE GRADE WITH ABRASIVE PLANING

By Sherman E. Kirchmeier

SHERMAN KIRCHMEIER, manager of Louisiana-Pacific's Oroville, California, division, worked as a log scaler while putting himself through Chico State College, from which he graduated in 1962. He became a scaler for F. M. Crawford Lumber Co. in 1956 and was promoted to forester for the Potter Valley mill early in 1966. He then moved to land manager and plant manager for the subsidiary Apache Lumber Co. mill before becoming plant manager at Aborigine Lumber Co. in 1968. After Georgia-Pacific Corp. bought both Crawford and Aborigine, Kirchmeier was transferred to G-P's Feather Falls mill on the Sierra side of California. There he was responsible for completion of the new mill and for closing the old operation. He was named manager of the Oroville division of Louisiana-Pacific at the beginning of 1973, upon the corporate splitting of G-P's system. Kirchmeier is a director for District VII of Western Wood Products Association and is on WWPA's technical committee. He is president of both the Oroville Chamber of Commerce and that community's industrial development corporation.

CHAPTER SUMMARY: The attention paid to abrasive planing is increasing in the sawmill industry as better wood utilization becomes synonymous with survival. Most mills now saw precisely enough to make abrasive planing worth considering.

There are two types of systems: batch and in-line. Louisiana-Pacific Corp. has installed the batch system at its Oroville, California, plant. It is a dual line with a pair of two-headed edge planers, two accumulating lines, and a four-head, 52-inch Timesavers abrasive planer.

Edge-planed boards fall onto the accumulator which aligns them for entry. Minimum feed can be 125 lineal feet per minute (or 1,625 board feet per minute).

Louisiana-Pacific has an in-line system at its Samoa, California, plant.

Boards are sanded four sides, one at a time, and profilers can be mixed with cutterheads. The in-line type is the easier system for converting an existing planing mill.

A mill requiring heavy stock removal at high speeds will not gain from in-line abrasive planing. Precision cutting and a need to avoid downgrade from chipping out or tearing of grain is most compatible with abrasive planing. However, Northwest Hardwoods, in Arlington, Washington, has handled thick stock removal at good production rates.

The Oroville mill cuts to closer tolerances without grade loss—often to 1/64 inch. The side head knife planers run 750 feet per minute and the abrasive planer 150—giving average runs of 240,000 feet per eight-hour shift.

Belt costs at Oroville have dropped from $0.50 to $0.75 per thousand board feet, in 1972, to $0.25 to $0.50 per thousand board feet in 1974. Belt changes are easy. 3M Co. is furnishing more durable abrasives which are cutting costs.

Louisiana-Pacific hopes to be able to finish its green lumber abrasively soon.

The main purpose of the Samoa mill is to size and finish finger-jointed stock. The Kimwood Corp. abrasive planer has two top and two bottom 15-inch heads. Before installation, 40 percent of glued stock was downgraded by tear-out or chipping. Belt costs at Samoa have dropped from $1.70 per thousand board feet to less than $0.25 per thousand board feet. There is no joint damage or grain tear.

Much remains to be learned about belt and feed speeds in abrasive planing.

CHAPTER 19 CRT DRYING FOR REDUCTION OF DEGRADE

By Dr. Dallas S. Dedrick

DR. DALLAS S. DEDRICK, a lumber drying consultant with Irvington-Moore, Portland, Oregon, is a native of Hobart, Oklahoma, and was educated there and in Iowa. Dr. Dedrick earned an A.B. degree in chemistry at Oklahoma City University and went on to earn his M.S. and Ph.D. degrees, in physical chemistry, at the State University of Iowa. He taught and also directed research programs at North Dakota State University, Iowa State University, and the University of Oregon. He joined Weyerhaeuser Co., in its

Research Division, in 1944, and from 1960 to 1970 was leader of the company's fundamental drying studies. After retiring from his position at Weyerhaeuser in 1970, he since has been a consultant with Irvington-Moore—although he continues an association with Weyerhaeuser as a consultant on problems in lumber drying. Among his professional affiliations, Dr. Dedrick counts the Forest Products Research Society, the American Chemical Society, the Society of Wood Science and Technology, and the American Society for Testing Materials' subcommittee on wood-water relationships.

CHAPTER SUMMARY: Continuously Rising Temperature (CRT) drying is a novel, patented process of heat transfer from a heat source to wood in a dry kiln. It is not a schedule. Constant heat transfer results in a constant rate of moisture removal.

In conventional drying, the rate of moisture removal decreases toward the end of the cycle. With CRT drying, kiln time can be cut sharply—reducing costs.

Much more important, however, CRT drying preserves grade far better than conventional schedules. About 3 percent of softwood lumber dried in the U.S. and Canada now uses CRT. Weyerhaeuser Co. reports an investment return of about 10 times the $1.4 million spent in development.

CRT-dried lumber is flatter and lighter in color than conventionally dried lumber. To date, statistically significant comparability studies have been made only with Douglas fir and southern pine dimension. On both, values were several dollars per thousand board feet higher than for conventional or high-temperature drying.

Sugar pine, however, has shown a severe brown stain.

Dry-bulb temperature should start its rise at the lowest possible level while drying fully across the load, and should end at the highest point the kiln will achieve or the wood withstand. Fast drying requires a large wet-bulb temperature depression, maintained by mixing a small amount of dilution air with the spent air at higher dry-bulb temperature.

It is preferable to use a specially designed kiln, but most existing kilns can be modified for CRT. Remodeling costs have ranged from $3,000 to $80,000 each. Factors determining the economics of remodeling versus purchasing new kilns include: (1) the amount of improvement required in kiln performance; (2) the physical characteristics of the existing kiln—heating, venting, air circulation, baffling; (3) the state of repair of the existing kiln; (4) the capability of the existing kiln to accommodate the required changes; (5) the cost of the changes; and (6) the anticipated low-maintenance life of the facility after modification.

Irvington-Moore offers an evaluation service on the cost of remodeling existing kilns and the probable performance of the kilns after modification.

CHAPTER 20 ULTRASONICS TO DETECT WOOD'S DEFECTS

By Erwin H. Bulgrin

ERWIN H. BULGRIN started out with Allis Chalmers Manufacturing Co. in 1936 and remained there until 1948, taking time out to serve in the U.S. Army during World War II. After completing his B.S. in forestry at the University of Michigan, he worked as a forester and logging and sawmill supervisor with the Oliver Iron Mining Division of U.S. Steel Corp. In 1957, he obtained his M.S. in forestry from the University of Michigan and worked as a project leader in log and tree grade research for the U.S. Forest Service Central States Forest Experiment Station. He transferred to the U.S. Forest Products Laboratory in 1960 and is now project leader in charge of quality and yield improvement in wood processing.

CHAPTER SUMMARY: With timber supply short, more substitution of competing materials for wood can be expected. But it is not a desirable solution. We need to improve our efficiency with wood.

Cutting decisions are based on what the workman sees. Inaccuracies in his information reduce conversion efficiency and profit.

Ultrasonics–sound having a frequency above the audibility limit of the human ear–show the most promise of revealing what is inside a piece of lumber. Sound is transmitted through the cell walls of the wood–which are equally dense in all species. Sound moves fastest vertically in the cell walls, more slowly radially, and still more slowly tangentially. Moisture is not a factor.

Therefore, irregularities such as compression wood, knots, and changes in slope of grain show up and can be visually displayed for a computer to read and evaluate.

Thus clear cuttings may be located.

Work at the Madison, Wisconsin, Forest Products Laboratory of the U.S. Forest Service thus far has concentrated on knot location. A lumber readout system may be developed by early 1976. Defects would be sensed and located, and the computer would evaluate and control the saws to recover the maximum yield. System cost is estimated at $100,000 and payoff at one year. It will be best suited for furniture, millwork, and molding plants.

A log readout system should follow in about two years.

Raising the lumber recovery factor is vital to profitable operation.

1

Municipal steam/power from wood/other waste

S. LaMont Matthews, Director of Mechanical Engineering,
Cornell, Howland, Hayes & Merryfield and Hill,
Corvallis, Oregon

The effective conversion of what we previously regarded as waste to energy has become dramatically more important—particularly in the early months of 1974. This chapter will discuss the results of investigation in the Lane County, Oregon, area of the feasibility of constructing a regional power plant for the disposal of excess wood waste and solid waste. The proposed plant would produce power and process steam for sale, thereby reducing costs of waste disposal.

The regional plant may offer an alternative to wood product mills, caught in the pinch between lack of fuel and a waste disposal problem. When a mill considers whether to construct its own power plant or develop other waste uses or disposal methods, the regional plant may prove to be a less costly "out."

There are two main reasons why the regional waste disposal/energy recovery plant is generating considerable interest. First, conventional disposal of solid wastes by landfill generally will not provide long-range solutions for communities larger than 5,000. Most communities of 50,000 or more are quite happy if they can locate an economical landfill with a projected life of seven years or longer. Second, we are painfully reminded every day that all forms of energy are in short supply. The possibility of solving the solid waste disposal problem, and at the same time producing steam and power, is attractive to the general public, as well as to industry.

Both steam and power produced

What type of plant is under consideration for regional waste disposal and power production? The plant proposed for the Eugene-Springfield area in Oregon is shown in Figure 1.1. Site size might be 130 acres. Each of four hogged fuel storage piles would be 700 feet in diameter and 50 feet high—for a total wood waste storage of 260,000 units. The proposed plant would have three wood waste-fired boilers, each producing 460,000 pounds of steam per hour. There also would be two refuse-fueled boilers producing 108,000 pounds per hour each. This would supply a 680,000-pound-per-hour process steam load and generate 115 megawatts of electrical energy. Total cost of the proposed power and disposal plant would be about $55 million.

Figure 1.1. Proposed regional waste disposal/power production plant for Lane County, Oregon. Site is 130 acres. Each hogged fuel storage pile is 700 feet in diameter and 50 feet high. In addition to process steam (680,000 pounds per hour) the plant would generate more power (115 megawatts) than nearby Detroit Dam (100 megawatts).

Projected waste availability

Figure 1.2 shows the volume of solid waste which would be produced in central Lane County over the life of the facility.

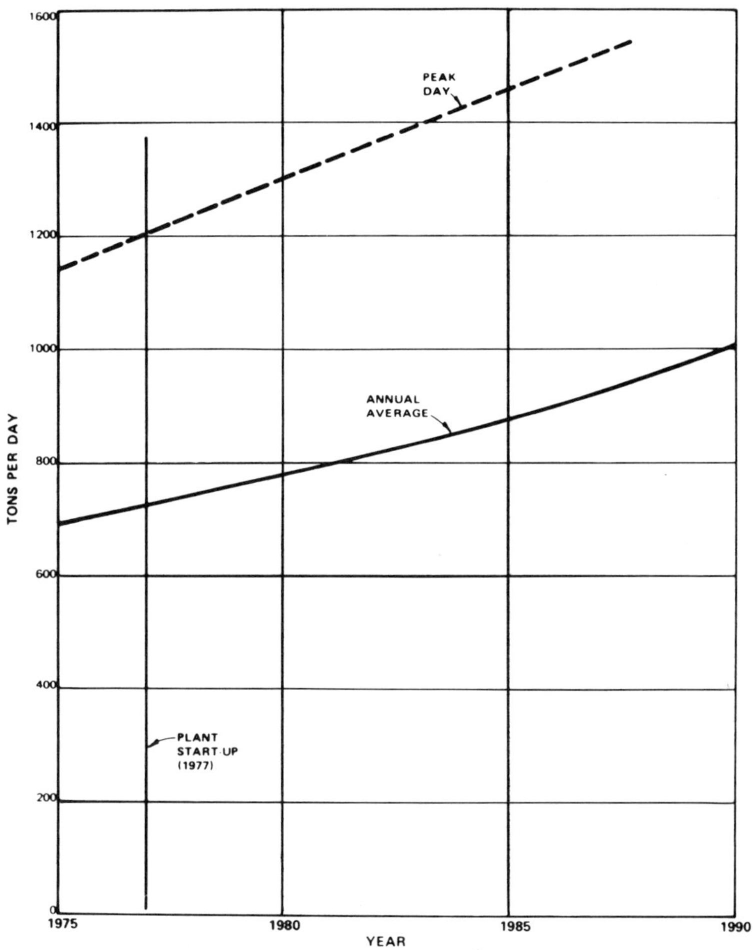

Figure 1.2. Volume of solid waste expected to be produced per day in the Eugene, Oregon, area ranges from 700 tons in 1975 to 1,000 tons by 1990. Curve follows projected population increase.

Starting with 700 tons per day and increasing to 1,000 tons per day, the curve reflects the projected population increase for the county. This solid waste is about 83 percent combustible—with 8 percent glass, 8 percent metals, and 1 percent dirt. Incidentally, after this material has been processed to remove metal and glass, it is essentially shredded paper and would make a fine supplementary fuel for sawmill or plywood plant hogged-fuel boilers.

Figure 1.3 shows amounts of wood wastes estimated to be available to this plant within a 20-mile radius and within a 40-mile radius. The data are based on a mill survey by the Eugene Water and Electric Board. Within a 20-mile radius, 1,600 units per day are available to start. By 1990, the amount decreases to about 1,500 units per day. This reduction reflects better log utilization and more uses being developed for waste.

With the energy crunch, the number of mills installing or expanding their own boilers is going to increase substantially. So some of the wood waste figures originally compiled may be open to question now. However, the basic energy recovery concepts remain. The relative economics of energy recovery systems, as compared with landfill alternatives, are still valid. In addition to wood waste and solid waste, this plant will take many additional wastes available in the county, such as sewage sludge or discarded tires. But the amounts of these materials are small compared with the quantities just discussed.

Investment quickly justified

Figure 1.4 shows estimated revenue produced over the life of the plant and compares it with revenue required to amortize the investment and pay for operation and maintenance of the facility. (All economic information is based on 1973 dollars.) After the first year, revenues would exceed the investment required to own and operate the plant. So the difference could be available to buy solid waste and wood waste as fuel. The revenue increase primarily reflects increases in the value of power over the 1975 to 1990 period.

Figure 1.5 shows how increasing revenue from power produced affects value of the solid waste and wood waste. For both waste categories, value in 1977 is very low, resulting in a charge of $0.16

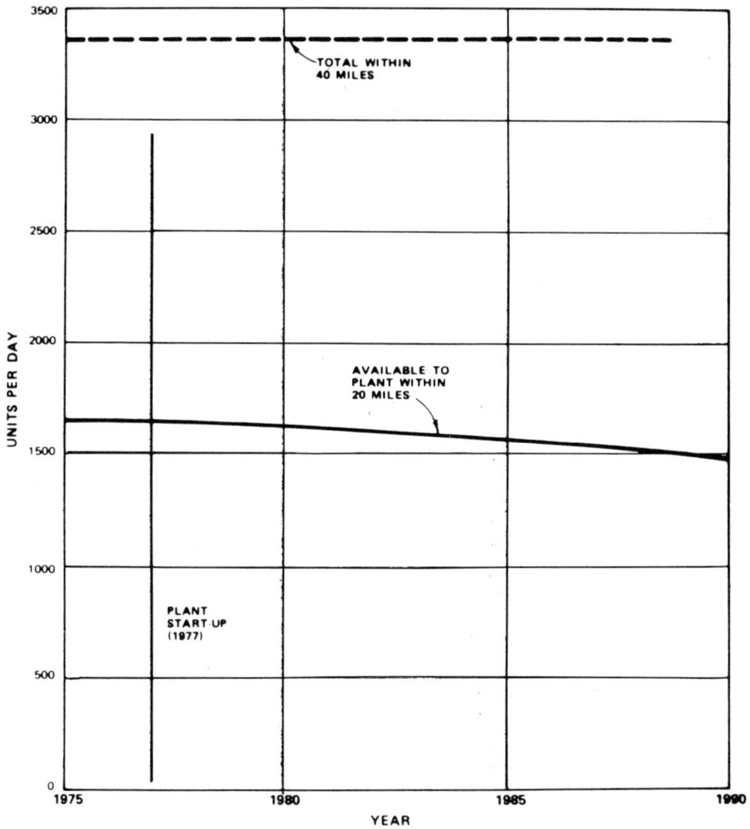

Figure 1.3. Volume of wood waste generated within both a 20- and a 40-mile radius of Eugene, Oregon. Volume within 20-mile radius is seen decreasing as better log utilization and more uses for wastes develop.

per unit at the plant gate to accept wood waste, and \$0.21 per ton at the plant gate to accept solid waste. However, by 1987 it would be possible to pay \$2.50 per unit for wood waste at the plant gate and \$1.07 per ton for solid waste. This does not include collection costs, so it really is not going to put the regional plant in the position of buying garbage directly from the householder, although it should reduce the householder's monthly garbage bill.

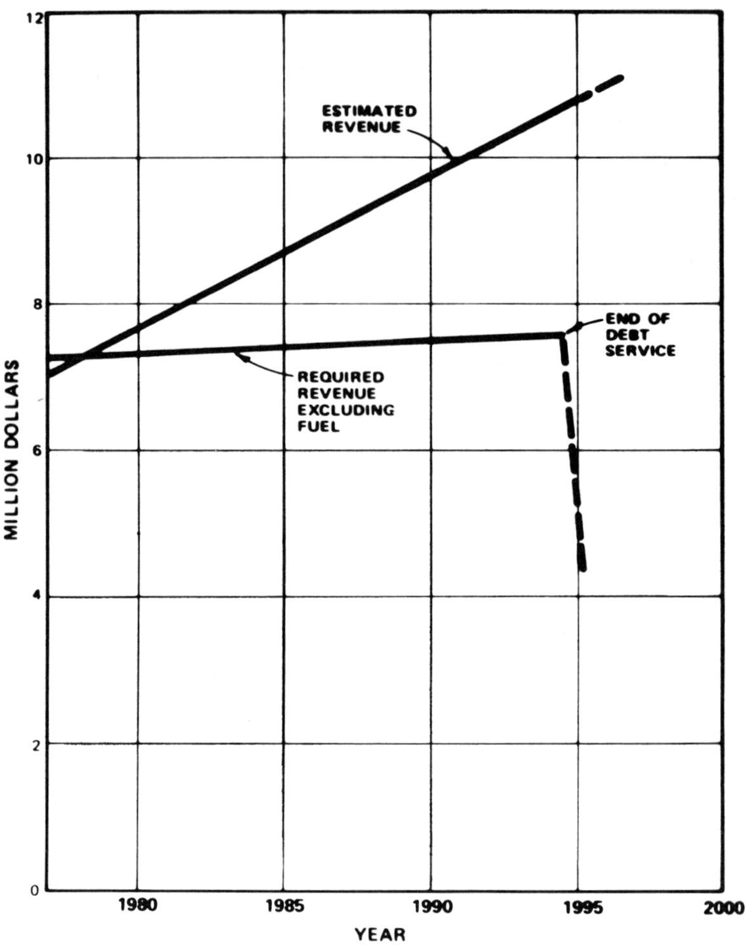

Figure 1.4. Revenue required to operate the proposed plant would exceed revenue produced only initially. Revenue produced will rise rapidly—with projected plant payoff date about 1994.

Making economics attractive

There are several ways to make the economics of this plant more attractive. One obvious way would be to defer construction for several years to take advantage of increasing electric rates. However, this may create complications in rapidly filling landfill sites.

Another possibility would be deferred financing of wood waste

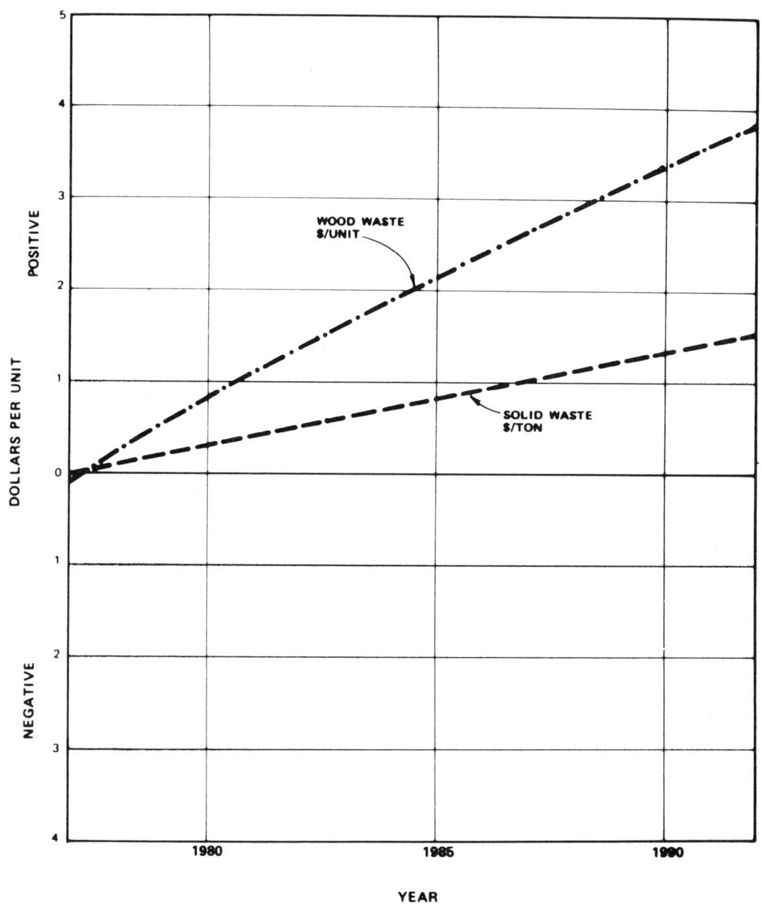

Figure 1.5. From having to charge for garbage and wood waste disposal early in its life, the operation would move quickly to being able to pay for them—making wood waste worth perhaps $2.50 per unit and reducing Eugene-Springfield residents' trash removal bills.

purchase. Figure 1.6 shows the effect of paying a minimum of $1 per unit for the hogged fuel. As part of the plant financing, this minimum value could be paid from the time the plant opens. This then would be reimbursed to the plant over a period of years as

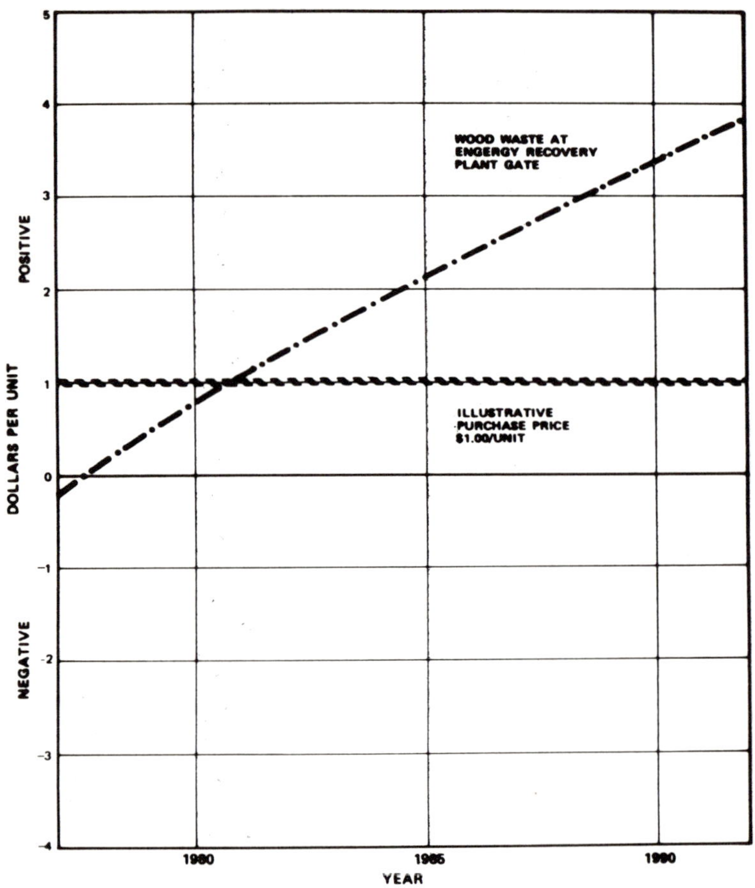

Figure 1.6. Plant operator could elect to pay $1 per unit for hogged wood waste to open the plant. This would represent an initial loss, but wood waste value would soon exceed $1 per unit, and the operation would begin to recover those costs.

the actual value of the wood waste begins to exceed $1 per unit—with full reimbursement occurring when the value per unit exceeds $2. In other words, as soon as the area under the horizontal line in Figure 1.6 equals the area over the line, the financing of the wood waste would be paid off. The operating agency could then begin to pay about $2 per unit for wood waste.

Figure 1.7 compares the cost of disposal of community solid waste through a plant using energy recovery with the cost of disposing of that same waste by landfill. In 1985, there is a difference of about $5.60 per ton. When a community produces 800 tons of this material per day, this represents a *difference* in disposal cost of $1.65 million per year.

Wood waste surveyed

Estimated wood waste available to this plant was based on surveys, as noted. Actual construction of a plant this size would require firm commitments from mill owners for a fuel supply. For a number of very valid reasons, it has not been possible to get these commitments. Some larger mills are talking about expanding boiler plants and about buying hogged fuel. Some medium-size mills may be able to justify boiler plant construction in 1974, due to fuel costs and shortages, when they could not have justified it six months earlier. So the quantity of wastewood for fuel has essentially become indeterminate during 1973-1974. There is, however, strong feeling among a number of individuals in the forest products industry that a significant part of the wastewood studied actually will be available.

Plant feasible without wood

However, if we consider the extreme case of *no* available wood waste, we still have the residential, industrial, and commercial refuse. A refuse only-fired electrical power plant for the Eugene-Springfield area would start with a generating capacity of about 35 megawatts. This reduced system concept would essentially break even during the first year of operation, with the economic return improving with time. The point is: economic feasibility of a waste-fueled energy recovery plant does not seem to be dependent

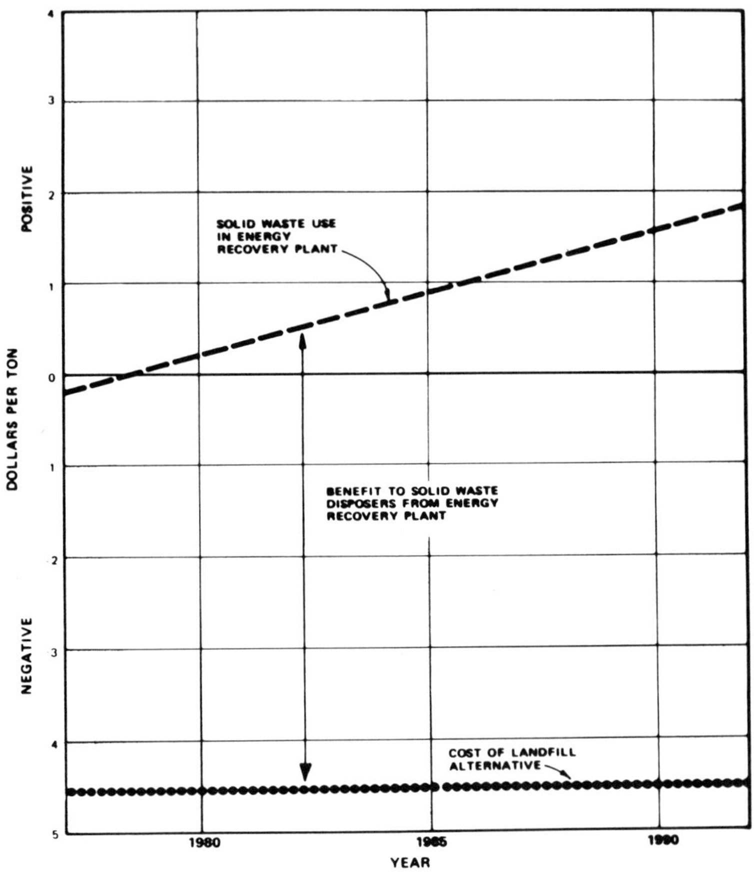

Figure 1.7. A community would be better off burning its solid waste to produce power—whether or not there is any wood waste in the mix. By 1985, a community producing 800 tons of solid waste per day would be $1.65 million better off per year by generating power from its waste than by burying that waste in a landfill.

on any minimum quantity of wastewood. Consequently, the first step ought to be the construction of a regional waste disposal plant to handle solid waste only. This concept is being pursued in the South Coast region, where a similar study is being conducted. Flexibility would be required—so that the facility could be expanded to include wastewood later. This would permit mills to consider the regional waste disposal plant as an alternative, along with industrial fuel sales and on-site boiler construction. This permits a publicly financed regional plant to provide a means of waste disposal—for both the general public and industry—in the most economical fashion possible.

2

Western potential for yield/grade improvement

H. C. "Carl" Mason, President, H. C. Mason & Associates, Inc., Gladstone, Oregon

We are all familiar with the prices now being paid for high-grade fir, hemlock, and pine stumpage. These prices are the product of competitive bidding on Forest Service and Bureau of Land Management sales and reflect very clearly the western lumber manufacturers' opinions as to what resource is worth competitively when applied against an identifiable market.

The Northwest enjoys a unique position in the world in being one of few areas capable of supporting sustained production of high-grade softwood clears and timbers of identifiable species in reliable and continuing grades and quantities. This wood commands a premium in the marketplace all over the world.

Those of you involved with export markets are certainly aware of the potential of and the demand of world markets for such fiber. Only woods such as lauan, mahogany, Virola-like species, and so forth, of the tropical or subtropical forests, have the potential of competing in these markets.

One thing that is not well understood, however, in the Northwest, is direct and indirect relationships between yield and grade and how those relationships can be improved.

New definition of yield needed

Obviously, yield and grade are two very separate and distinct items and are not always compatible. First, yield, as expressed in the West in overrun Scribner log scale to lumber-board-foot scale, should be modified to express recovery in terms of the lumber recovery factor (LRF), which is expressed as board feet of wood product recovered from a cubic foot of log input. This is, obviously, a much clearer definition of recovery than is the conventional method of overrun based on net Scribner log scale.

All of us realize the variability of wood recovery, either in terms of lumber recovery factor or overrun, as it applies to the different products manufactured; that is, the substantial variance between a scant-sawn 2x4 (which requires approximately 72 cubic feet of wood to produce 1,000 board feet of salable product) and full-sawn timbers, or full-cut or full-sawn selects, and export (which require from 100 to 120 cubic feet to produce 1,000 board feet of salable product) due in part to the measuring system used in the United States. In most cases, timbers, uppers, and export are full sawn or, in some cases, oversawn in order to accommodate grade requirements.

Therefore, without becoming involved in the measuring system per se, I would like to discuss the relative improvements which can be made to the manufacturing of lumber, both dimension and uppers, to increase recovery from the raw material.

Factors determining yield

Yield, or recovery, is generally the product of four things:

1. Product line or the types of products produced: full sawn versus scant sawn, and so forth.
2. Oversawing to accommodate variations in sawing.
3. Kerf, planing, and shrinkage allowances.
4. Method of breakdown or technique employed in sawing the log into lumber products.

Product line. Product line is the best mix of products for a given grade and species to produce the maximum return.

Sawing variation. Sawing variation pertains to the allowances that are made in the system in terms of variation to produce a product acceptable in the marketplace or, in other words, that variation around the standard established as minimum. A good example is dimension lumber, which in many places in the Northwest is cut on the basis of approximately 1 5/8 inches (from 1.625 to 1.70), although various mills cut acceptable lumber within the range of 1.650 to 1.675. Obviously, the size that can be cut depends entirely on the accuracy of the machine tools in the mill. Principally, it is determined by the accuracy of the most inaccurate major piece of machinery on the mill floor. Therefore, if all machines can operate within a sawing variation of 0.060 inch and the minimum-size dimension lumber allowed for adequate planing, adequate surfacing, and drying is 1.620, then the set sizes would have to be at least 1.680, or the sum of the two, 0.060 and 1.620.

Allowances for processing losses. Machine allowance, the allowance calculated for shrinkage, planing allowance, and kerf are the other mechanical variables that account for the balance of the loss in producing lumber to dry finish sizes. They play proportionate parts in the relationship of yield to recovery. For example: lumber being finished to 1.500, requiring 0.060 for shrinkage, plus 0.060 for surfacing, plus 0.060 for sawing variation would require a 1.680-inch rough green size. If kerf were 0.120, the gross wood involved in producing a piece of dimension lumber would be 1.800. Accordingly, a 2x6 cut 1.800 by 5.800 would require 72.5 cubic feet of wood, including lumber shrinkage, planing, and sawing variation and kerf to produce 1,000 board feet of salable product.

Method of sawing. An additional determinant of wood recovery is the method of sawing. There is substantial difference in yield among logs that are live sawn, half-taper sawn, full-taper sawn, symmetrically sawn, quarter sawn, or B.O.F.-handled. Consideration of the method used to convert logs to lumber should enjoy high priority. Actually, the potential available to the manufacturer in scant precision-sawing on high-grade material is substantially more valuable than it is on lower value dimension and commons. On the whole, I believe the industry is tuned in to the advantages of cutting *dimension* lumber as accurately, as simply, as scantily as possible and is attempting to do so. However, this philosophy has not been very well accepted or practiced in the cutting of uppers, clears, export, and timbers. It is these high-grade products, which command very high prices in the marketplace, that have great potential on the West Coast to increase mill profits.

Recovery and grade of lumber products are generally determined by the quality of the log and the type of machinery used to break down that log into lumber. The difference among breakdown techniques for similar end products is profound. As an example, Figure 2.1 describes and compares some widely used methods of log breakdown.

How cutting factors determine recovery

Generally, live sawing (with logs relatively defect-free and recovery the only consideration) will show the highest recovery when the log is cut into dimension products. However, we all know that knot patterns in the center of the log, producing spike knots,

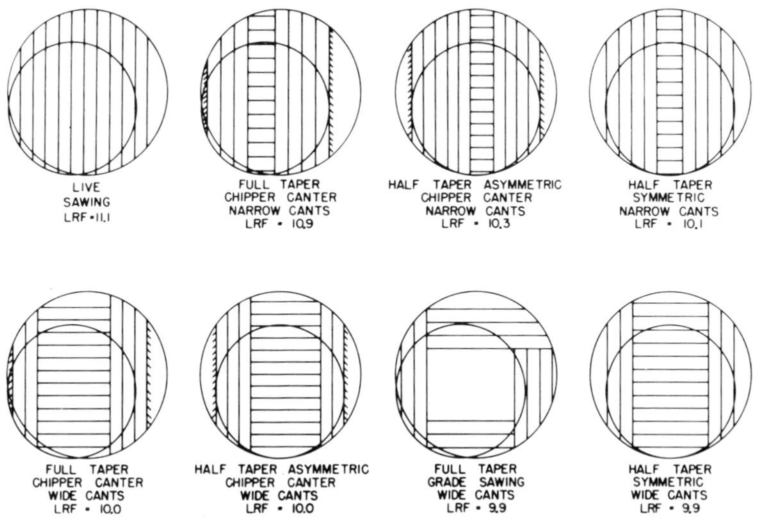

Figure 2.1. Various log sawing or breakdown methods for an 18-inch-diameter log 16 feet in length. In each case, larger circle represents large end of the log; smaller circle represents small end.

often dictate cutting patterns that avoid live sawing through the heart. The point I wish to make is that there is a substantial difference in recovery based on how you cut: how you cut is based on the combined analyses of the species and type and size of log, together with the end products that are desired. All these factors have a profound effect on recovery. Figure 2.2 shows the amount of wood required to produce 1,000 board feet of salable product for a number of lumber sizes.

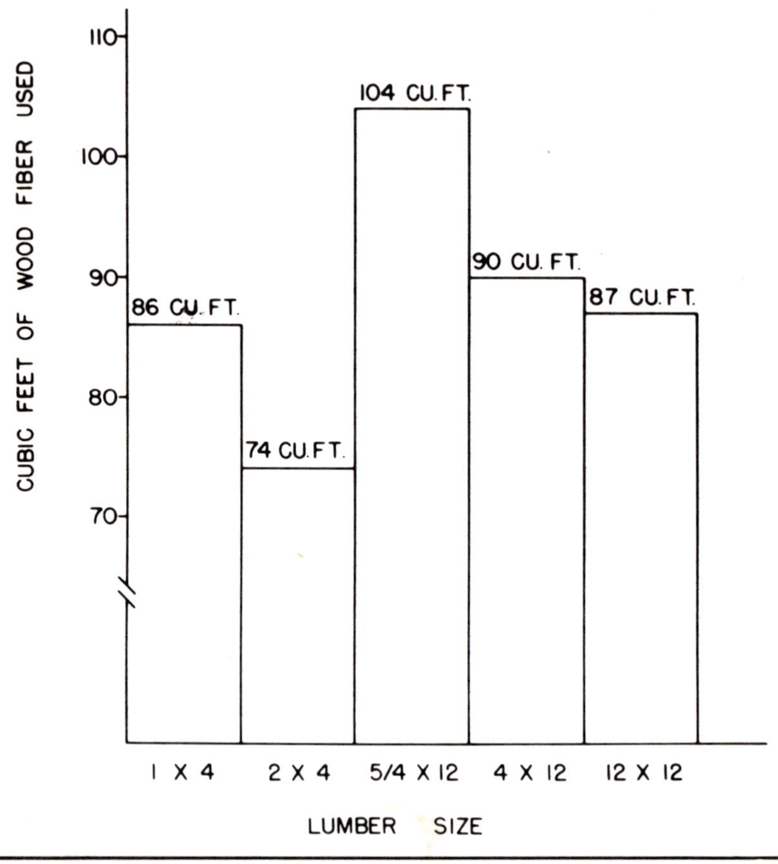

Figure 2.2. Amount of wood fiber required to produce 1,000 board feet of lumber varies with the product recovered.

The fundamental logic that is required to convert this type of analysis into dollars is the acknowledgment that every ounce of fiber that is saved by reduced kerf, less shrinkage, less planing allowance, less sawing variation, or better geometry from cutting patterns will result in additional end product. This fact was not well understood a few years ago, but I believe it is much better understood today. Figure 2.3 shows a typical analysis of the relationship between fiber saved and increased end product.

Diameter a determinant of recovery increase

A 4-inch diameter log which can produce only one 2x4 is not affected by any of these factors; therefore, any improvement in kerf, sawing patterns, and so forth, would not increase the end product.

Accordingly, at approximately 4 inches in diameter the rate of return, due to fiber savings, is zero. But, if a log were of infinite diameter, surely anything recoverable would end up in additional end product. Therefore, a curve between a log of infinite diameter

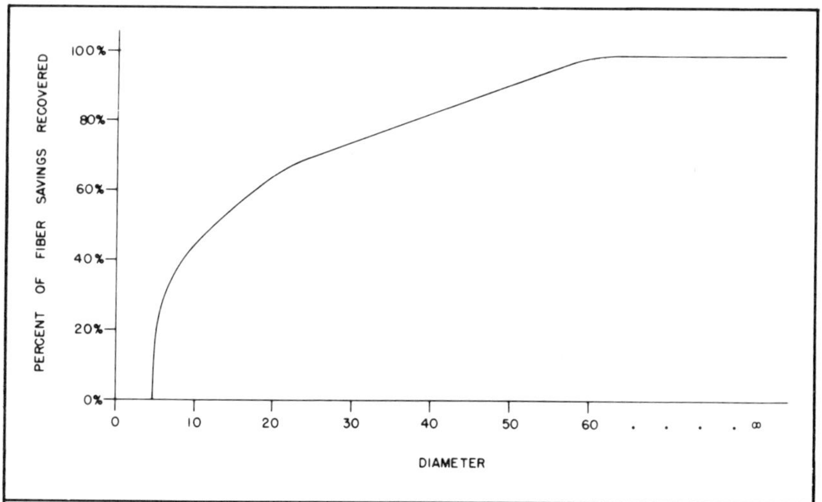

Figure 2.3. Percentage of wood which can be saved (or lost) by better or worse sawing techniques. Gain in end product recovery is much greater in the larger log diameters.

and one 4 inches in diameter will show a relatively high rate of return for diameters in excess of 16 inches and a rate of return well in excess of 50 percent for logs even as small as 10 or 12 inches in diameter.

Obviously, this curve would be different for each mill. A curve for a mill that has a relatively high recovery factor with thin kerf and little wood loss would be very different from the potential for a mill which has large kerf, poor sawing practices, and, obviously, much more room for improvement. My point is, all savings can be projected as a predictable and identifiable rate of increase in end product recovered from the raw material.

Analyzing mill functions for profitability

All these functions can be modeled, simulated arithmetically, and computer tested. There are a variety of techniques for making these analyses—with a high level of reliability. The reliability of the analyses is based on two premises: (1) that a major portion of all calculated savings can be found in additional end product or can be programmed into additional end products, and (2) that any mill operating competitively today is run in an orderly, business-like fashion and its decisions are based on intelligent rationale. By determining kerfs, sawing allowance, fiber loss, and so forth, and by determining the amount of fiber used to develop 1,000 board feet of salable product in any mill, any reasonable, hypothetical alternative can be applied with a very high degree of predictability. When this type of analysis determines a significant potential increase in recovery, based on change of product, change in kerf, or change in sawing practice, we can make these changes and see the results in additional product. We have been doing this in our business now for some six years, over a wide range of mills throughout the United States, and are amazed at the reliability of the results obtained by this type of analysis.

How to improve lumber grade

I would like to point out how grade may be improved in the West. Grade cutting is primarily the product of the resource and occurs only in old-growth stands of fir, hemlock, pine, redwood, spruce, larch, and cedar. The techniques mentioned in volume recovery, or yield recovery, are applicable to grade recovery. However, the optimum recovery techniques for yield very often are incompatible with grade. They then must be traded off or compared on the basis of end profit dollars to determine the best solution. A high average selling price and a low rate of wood recovery may result in a poorer profit margin than does a lower selling price and a higher recovery, again depending on the grade and quality of log.

When you accept the fact that any fiber saved is identifiable in additional end product, you can put a very real value on this fiber. Referring to the yield curve previously discussed, we found that the rate of return increased with the diameter of the log. We also know that the majority of logs cut into clears, uppers, and timbers are logs in excess of 24 inches, where the rate of return is generally in excess of 80 to 85 percent of that calculated.

Improving grade involves a variety of techniques–basically, mill floor design and the equipment employed in cutting logs. In conventional systems complete breakdown on bandsaw and carriage provided the operator with maximum flexibility in selection of grade and employed the highest skill of the sawyer. But there have been several deterrents to this technique: (1) capital investment, (2) kerfs employed on headrigs, (3) the general inaccuracies of such things as carriages and headrig bandsaws (caused by the high level of impact shock and abuse on this equipment) and their limited production.

The past five years have produced improved downstream mill equipment so that saw lines can be kept to a minimum on the headrig, with a maximum of cutting done by thinner kerf, higher precision equipment downstream, primarily resaws or pony rigs. In addition, it is very difficult to identify what the sawyer intended in the original breakdown for sections of logs after they have reached the back of a mill. So most techniques for downstream breakdown often have had to compromise grade for recovery.

New breakdown techniques and equipment

Today, I do not think this is true. Since 1970 a number of studies and much developmental work has related to development of sawing patterns or techniques to break down a log with minimum saw lines at the carriage. This involves precision, high-yield equipment downstream to do the balance of the breakdown. These techniques provide the ability and flexibility to attain increased yield without sacrificing grade. The most useful tools available are the precision, solid linebar, high-strain twins with precision setworks and thin kerfs. These resaws are capable of handling log sections as large as 30 inches deep and 48 inches wide and, in effect, able to do everything that a headrig can do once two sides have been cut on the piece at a 90-degree angle to each other. Figure 2.4 shows several examples of this technique, which provides a flat for conveying and one straight edge for the linebar; this provides the operator with a stable platform and, in effect, allows the operator most options available to a sawyer with a bandmill and carriage.

Another major equipment development is the computer-controlled, precision ball-screw carriage, which has a high level of repeatable accuracy. This equipment also minimizes the mechanical work load imposed on the sawyer—by removing the carriage dial and providing the sawyer with a digital approach. These systems require the sawyer to think only in nominals—4/4, 6/4, dimension 4-, 6-, 8-, 10-, 12-inch, and so forth. Only the computer knows exactly the set size—but it knows this to three decimal points. Accumulation of multiple lines, backstands with a width-size cant, plus any number of nominal lines, are all calculated by the computer, releasing the sawyer to concentrate on log handling and decisions as to products to be cut.

The use of the computer as a control mechanism, the precision ball screw as a means of adjustment and measurement, absolute encoders, and prearranged sawing geometry is providing an entirely new level of performance and flexibility of techniques never before available in the wood industry.

Profitability does cost

Profitability in the lumber business is the result of making the right products from each log, generating the greatest amount of salable product possible, and doing it at the least possible cost. The rules for accomplishing this have changed greatly in the past

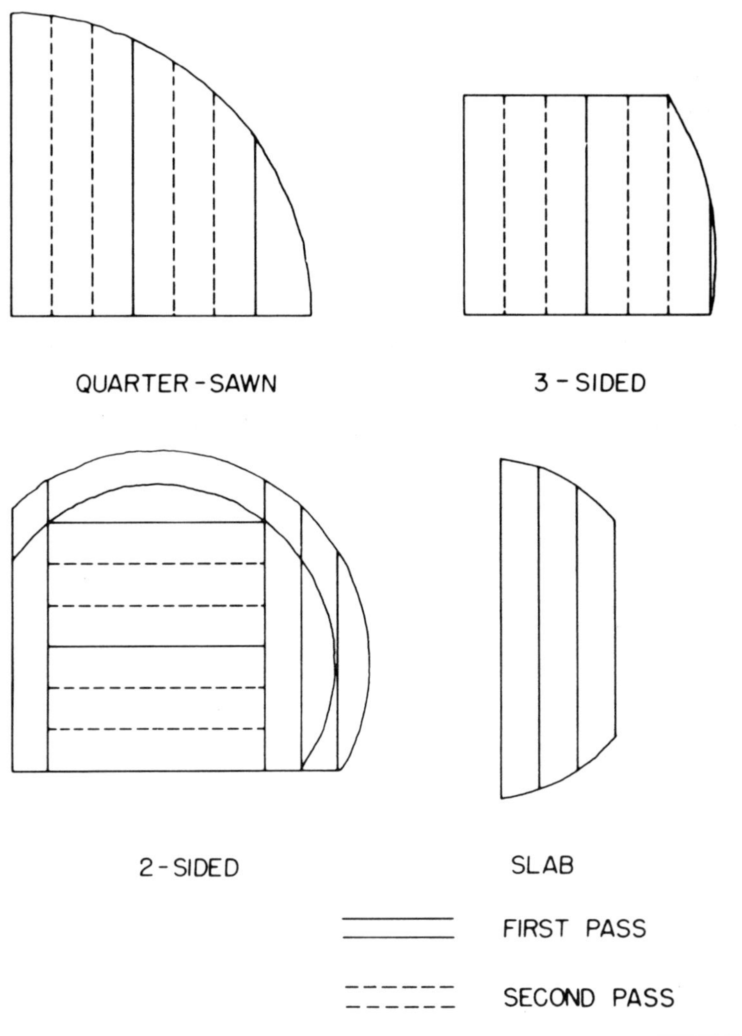

Figure 2.4. Examples of different sawing patterns for a twin-band resaw. Once two sides have been cut at 90 degrees to each other, the sawyer has most of the options available to a headrig, while enjoying high precision and narrower kerf than on a carriage headrig.

five years. Production rate is no longer a sacred cow. A cheap, haywire mill cannot compete. Mill operators, supervisors, and machine operators cannot be left to make decisions based on past experience. The industry today is highly competitive and capital-intensive. A sawmill carriage costs one-quarter of a million dollars; a simple bandmill, carriage, track, deck, and feed cost in excess of $1 million, but it will cut lumber accurately to a few one-thousandths of an inch.

Importance of raw material costs

We are currently paying nearly $300 per thousand for timber on the stump. Add $40 or $50 logging and hauling costs, and $350 logs at the mill are a valuable commodity. They should be treated carefully and sawn slowly, with every consideration given to both grade and recovery. Timber is valuable material. The capital cost of the equipment to convert logs to lumber is a fundamental part of unit cost; the number of people involved also is fundamental to unit cost—but both are secondary to log cost. In an industry where raw material constitutes 50 to 80 percent of total production cost, or product value, then our attention must be directed not exclusively to production, not exclusively to cost of facilities, not exclusively to cost of manpower, but to the relationship which each of these cost items, including logs, have to total cost. Figure 2.5 graphs component costs of lumber in relation to selling price for the period from 1956 through 1973.

As an industry, we are faced with increasing unit costs due to increases in labor, capital cost, supplies, energy, and a substantially reduced production rate. At the same time, increased conversion costs are overshadowed by the more rapidly increasing value of raw material. The world is entering the age of shortages. We are just beginning to feel and identify with these changes. They will continue. Wood, which has been a commodity badly treated in our economic system of supply and demand, will eventually assume its rightful place as our only reproducible resource.

Opportunity knocks at industry's door

All types of paper are in critical supply. Most of the world is out of toilet paper. Publishing businesses are severely curtailed by lack of stock while demand increases. The energy crisis is probably

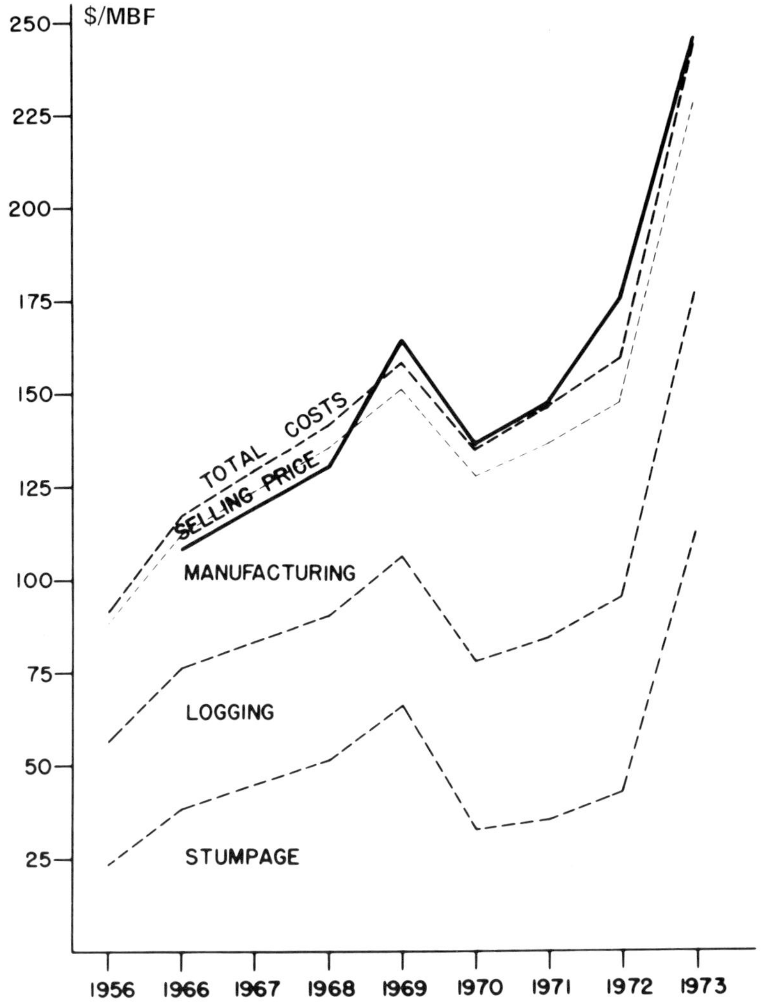

Figure 2.5. Component costs of lumber in relation to selling price for the period 1956 through 1973. (Based on U.S. Forest Service data for Region 6, Zone 2.)

here to stay. Because of the priority for petroleum products, polyvinyl chloride moldings will increase in price and be limited in quantity, thus increasing the opportunity for wood moldings. The demand for railroad ties is great and will increase as the American railroad system rebuilds—a real opportunity. Wood, as fuel, has and will take on new dimensions. Fireplace wood, at $50 to $70 per cord, is competitive with many products. Oil, at $15 a barrel, compares with wood at $18 a green ton as a source of fuel for steam and electricity. Eventually, all sawdust and shavings will be used as furnish for pulp and paper. Particleboard will have a difficult time competing for that fiber.

All these factors spell opportunity for the forest products industry. We must protect the resource. Fight to keep all forest lands productive and increase their productivity—including private, state, and federal forests. Practice quality control, slow down, increase yield, look for grade and premium products, keep up with changes in technology and markets. This is a rapidly changing time we live in, with unequaled opportunity.

3

Drying hogged fuel for powerhouse operation

Harry J. Hardy, Engineering Supervisor, Canadian White Pine Division, MacMillan Bloedel Ltd., Vancouver, British Columbia

This presentation concerns the hogged fuel preparation system presently under construction at the Canadian White Pine (CWP) Division sawmill of MacMillan Bloedel Ltd., at Vancouver, British Columbia. Purpose of the installation is to reduce fly ash emissions from the powerhouse stack; this will be accomplished by a self-sustaining hogged fuel drying system. Within the company, two other powerhouse emission abatement systems now are operative. Each attacks the problem after the fly ash has formed. The CWP system is designed to prepare the fuel for burning—to minimize formation of fly ash. The system is expected to reduce emissions to the Vancouver Pollution Control Branch level B, which is 0.15 grains per standard cubic foot of flue gas.

Hogged bark moisture reduction methods

At Canadian White Pine we produce about 1 million board feet of lumber every 24 hours. Production is divided roughly 50-50 between western red cedar and hemlock. Cedar hogged fuel is both the wetter and the harder fuel to handle. This presentation uses the cedar moisture content only to avoid confusing the two.

As discharged from the hog, hogged bark is 70 percent moisture, wet basis. Or, for every pound of bone-dry bark, there are 2.3 pounds of water. In 1970 we installed a bark press at a cost of about $250,000. With this, we reduced moisture content to 60 percent; or, with every pound of bark, 1.5 pounds of water. The $1.2 million dryer now being installed should reduce the moisture content to 44 percent, resulting in 0.8 pound of water per pound of bark. You will note that it cost $250,000 to reduce the hogged bark moisture content from 70 percent to 60 percent (taking out 34.7 percent of the water) but that it will cost $1.2 million to remove an additional 30 percent of the water. The reason: the first 34.7 percent was external water, which could be squeezed off by the bark press; the next 30 percent is internal water, which must be evaporated by heat. To accomplish this, the fuel will be prepared in two stages: (1) by secondary hogging to reduce the raw hogged fuel to 3/4-inch-minus particles, and (2) by reducing moisture content from 60 percent, wet basis, to 44 percent by drying in a rotary dryer.

The combination of finer and drier fuel will result in more complete combustion with less fly ash. It will also result in improvements in boiler operation:

1. Stable furnace conditions and a hotter furnace.
2. Reduced velocity through the furnace, resulting in longer retention time for fuel particles.
3. Reduced excess air.
4. Lower undergrate air pressure and better overgrate air fuel mixing.
5. Less efficiency loss from superheated steam in the stack gases.
6. Reduced gas volume at furnace exit, providing lower gas velocities in boiler banks.
7. Less boiler bank erosion.

Operation of the system

This is how the system works. (Figure 3.1 shows flow patterns and equipment in the system.) Hogged fuel comes from the mill, through the bark press, up a conveyor, across the top of the powerhouse, and into distributing conveyors over the tops of the furnaces. Excess fuel goes around the circuit back into the hogged fuel pit, where it joins hogged fuel coming from our Vancouver Plywood Division. A bulldozer moves the mixture of fir bark and recycled bark into the reclaim conveyor. It moves up that conveyor, joining fuel from the pit, and on through the system.

How the powerhouse is supplied with fuel

The hog reduces the size of the fuel and meters it to the system. We have installed a new 60-inch belt conveyor which passes a metal detector and then moves on through a Gruendler hog that will refine the material to 3/4-inch-minus particles. Material then moves into a 35-unit Clark metering bin, where it will be fed out by a metering screw at the bottom of the bin into a chain conveyor system at the rate of 63 tons per hour. The material—made up of 27 tons of bone-dry fuel and 36 tons of water—moves by conveyor into the powerhouse.

Drying of the hogged fuel

For drying, the hogged fuel passes over a 30-inch-diameter air-lock metering screw feeding the dryer. The metering screw keeps the dryer full and the excess passes over and goes on into the powerhouse undried. Hogged fuel enters a 9½-foot-diameter by 60-foot-long Sterns Roger rotating dryer. Temperature of the infeed is 1,200 to 1,600 degrees Fahrenheit. At the discharge end of the dryer, temperature is 250 degrees. A 200-horsepower blower pulls about 68,000 cubic feet of drying gases through the dryer per minute. They are made up of 13 percent water vapor, 37 percent combustion products, 10 percent infiltration air, and 40 percent recirculating gas. These drying gases will extract about 31,000 pounds of water per hour from the hogged fuel and discharge it through the stacks to the atmosphere.

How salt content is handled

An unavoidable component of our hogged fuel is approximately 2½ tons of salt per day. The high discharge temperature of about

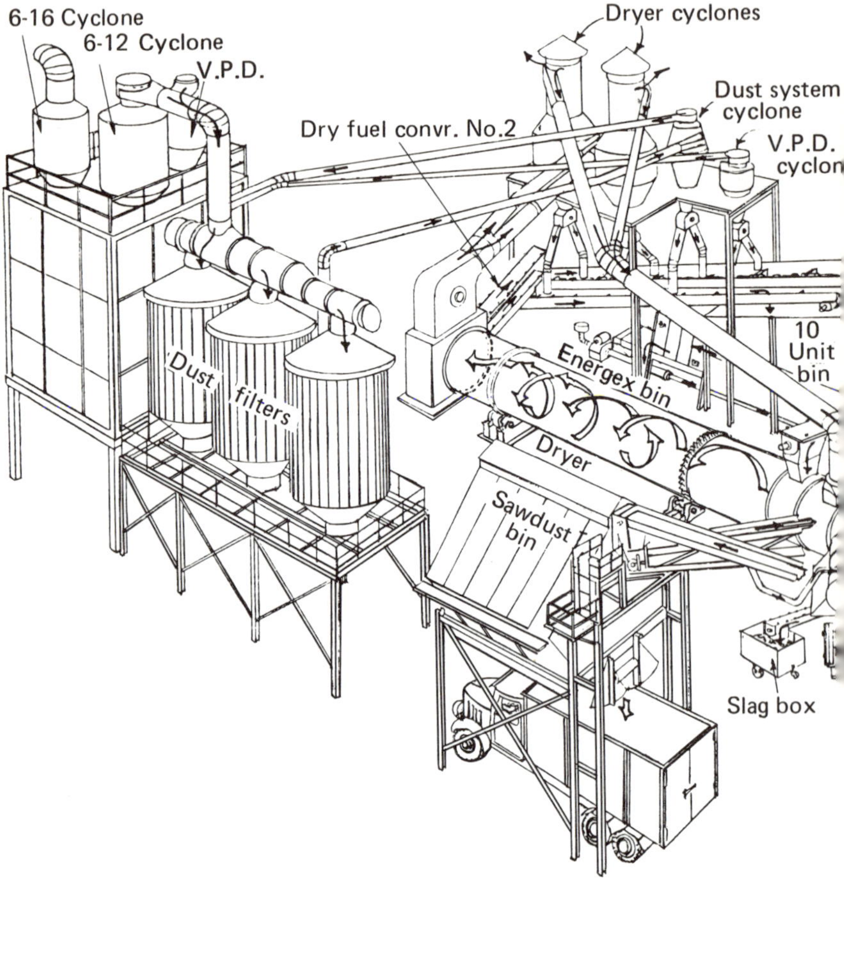

Figure 3.1. Equipment and flow patterns used in preparing hogged fuel mill in Vancouver, British Columbia. This sawmill is one of the world's larg

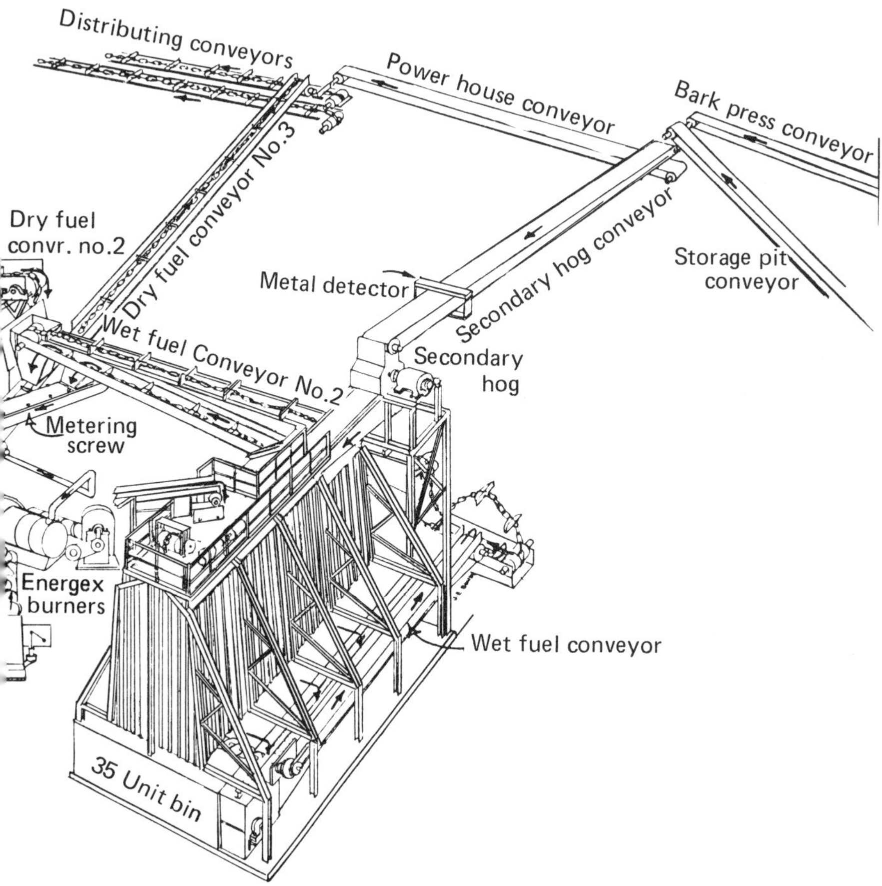

›werhouse operation at MacMillan Bloedel Ltd.'s Canadian White Pine Division th an annual production of 250 million board feet of lumber.

250 degrees Fahrenheit insures a stack temperature above the salt dew point. We will maintain an exit temperature as low as possible for drying efficiency but high enough to avoid condensation of salt in the duct. The discharge temperature also controls the Energex heat output. Furnace pressure controls the gas flow through the dryer drum, which is the transport medium, as well as the damper in the fan. The drum differential is the measure of dryer load. It controls the furnace input rate via a direct-current-drive, 30-inch-diameter metering screw which must maintain an air lock at all times to avoid extra air in the system. A 30-inch-diameter air-lock screw discharges the processed fuel into the chain conveyor system, which transports it to the powerhouse.

All chain conveyors have high-molecular-plastic wear strips to reduce the squeal anticipated under dry fuel conditions. Conveyors are covered and evacuated with a dust collection system.

Collection and distribution of fines

An eight-mesh screen is installed at the discharge end of the dryer (the last 42 inches) to extract a portion of the fines from the hogged fuel. Fines are conveyed by a 12-inch-diameter air-lock screw into an open-top chain conveyor, which distributes them into a 10-unit Lamb storage bin. Fines also are collected from the drying gases by the two 11½-foot-diameter cyclones and are discharged via tail pipe switching into the fines conveyor, or into the hogged fuel conveyor if the fines bin is full. A third source of fines is Vancouver Plywood Division sander dust, which is collected in a cyclone and discharged to either the fines conveyor or the hogged fuel conveyor via tail pipe switching. All fines end up in the Lamb 10-unit storage bin.

The last source of Energex fuel is dust collected from the planer mill cyclones and the hogged fuel drying and dust collecting system by three Clark filters. The screws at the bottom of the Lamb storage bin charge the Energex 100-cubic-foot metering bins, which (through a high-pressure blowing system that moves the material through the pipes) supply the two 27.5-million-Btu-per-hour Energex burners.

An Energex is a 36-inch-diameter by 60-inch-long cyclonic burner into which we inject air and fine dry particles in the right mixture to create optimum combustion. Its interior temperature is 2,700 degrees Fahrenheit, which is just above the melting point of ash.

In addition to salt, another unavoidable component of our hogged fuel is approximately 10 tons of sand per day, representing 31 percent of our emission. We will recover as much loose sand as possible through the rotating screen at the discharge end of the dryer. Recovered sand will be blown back into the Energex burners, where it will be melted and discharged through the slag tube in the bottom of the Energex burner into the slag recovery system. Slag drops into a screw bottom water tank and is conveyed by the screw conveyor into a disposal box.

This combination of refining, drying, and partially cleaning the raw hogged fuel before injecting it into the boiler furnace is how we have elected to reduce our emissions to meet the Provincial and Vancouver Pollution Control Branch requirement of 0.15 grains per standard cubic foot of gas.

We would be happy to have interested individuals visit the CWP Division Vancouver installation to learn about our system.

4

Briquetting wood waste by the Hausmann method

Fred Hausmann, President, Fred Hausmann Ltd., Basel, Switzerland

"Sam, take care that the people work! But don't ask anybody to work for nothing." These words are a hundred years old and still belong today to the most treasured quotations of American business; they were first pronounced by a man known to everybody in this room.

He was a teetotaler, nonsmoker, and apparently religious. At the age of 24, he was a grocery wholesale dealer. By 31, he owned the most important oil refinery in the United States. He was the son of a quack and hated extravagances. In his business methods he was neither squeamish nor sentimental. More than 100 years ago, this man founded the Standard Oil Co. And about 100 years ago, he originated the first oil boycott in history.

On January 2, 1872, he met in New York with the chairmen of the three then most important railroad companies. It was agreed to triple the freight rates for oil. But this man and a few other refinery owners would get a kickback of up to 50 percent on the new rates. Thus Standard Oil would profit from every carload of oil shipped by its competitors. (As I said before, this man was neither squeamish nor sentimental in his business methods.)

The planned cartel exploded, as the secret instructions for the new rates were left by mistake on a desk at the Lakeshore Railroad Co. An employee read them and let the story out. Thereupon oil producers cut back production by one-third and the first oil boycott in history was underway.

In the meantime, you have guessed this man's name—John Davison Rockefeller. He is a symbol for success, wealth, and power—for every boy, not only in the United States, but also in small Switzerland, my home.

But what has John D. Rockefeller to do with briquetting of wood waste? Quite a lot!

Origins of briquetting wood waste

The idea of briquetting wood waste is not new. Briquetting had been done—with more or less success—in the 1920s. The idea was derived from the wish to make use of all waste, rather than throw it away. And it is here that we meet John D. Rockefeller. In his planning and calculations he economized to the last possible cent.

He made money out of waste—a large amount of money. In production of petroleum, Rockefeller made everything into money. What his competitors were throwing away, as useless waste, he used. For instance, residues of the distillation process, which to his competitors seemed useless, Rockefeller converted into lubricating oil.

Waste has always existed in our society of plenty—always and everywhere. We have ignored it, thrown it away, or burned it uselessly.

If John D. Rockefeller were alive today, he would with certainty be here in my place. He would tell you how much oil he saves daily by making use of wood waste.

As I mentioned previously, the idea of briquetting wood waste is not unique. The following appeared in a 1923 publication:

> The only installation in the United States today for the briquetting of wood shavings for fuel purposes is in Los Angeles, California. The Pacific Coal & Wood Co. is placing the fuel upon the market. The wood waste is briquetted in a special brick press. The pressure ob-

tained is comparatively light. The briquettes are cylindrical in shape, about 3 inches in diameter and 10 inches high, and are finally bound with baling wire to insure their retaining coherence under shipping strains. Though not subjected to extreme pressures, this briquette is reported to be an excellent fuel, the only criticism being that the wire in the ash causes some difficulty in ash disposal.

In Germany, Austria, and even in Switzerland, installations for briquetting wood waste existed more than 50 years ago. But both in the United States and in Europe, these installations had little success and were shut down.

Necessity fostered utilization

Fortunately, in times of war mankind turns resourceful. During World War I, application was made for a patent for a "press for manufacturing logs from damp paper, etc." Figure 4.1 shows the official Swiss patent specification and a drawing of the patented press.

During World War II, the idea of briquetting was taken up again in Switzerland in a rather rudimentary way. Between 1939 and 1945 nearly every household owned its little briquetting press, which consisted of a container, a pressure plate, and a lengthened lever arm (Figure 4.2). Newsprint, soaked in water, was used as raw material. The container had small holes at the bottom and sides, allowing the water to be squeezed out. The log produced had an oblong shape. Briquetting, as a rule, was done by the lady of the house. Drying was left to the sun. We used such a press at home.

The press shown in Figure 4.3 was designed in Switzerland in 1942. Patents were taken out in Switzerland and England. This machine was the forerunner of today's mechanical extruder.

The difference between the presses of the 1920s, which in part still survive, and today's Glomera or Fred Hausmann presses is insignificant in terms of operational sequence. From a technical point of view, particularly regarding the design, progress has been remarkable.

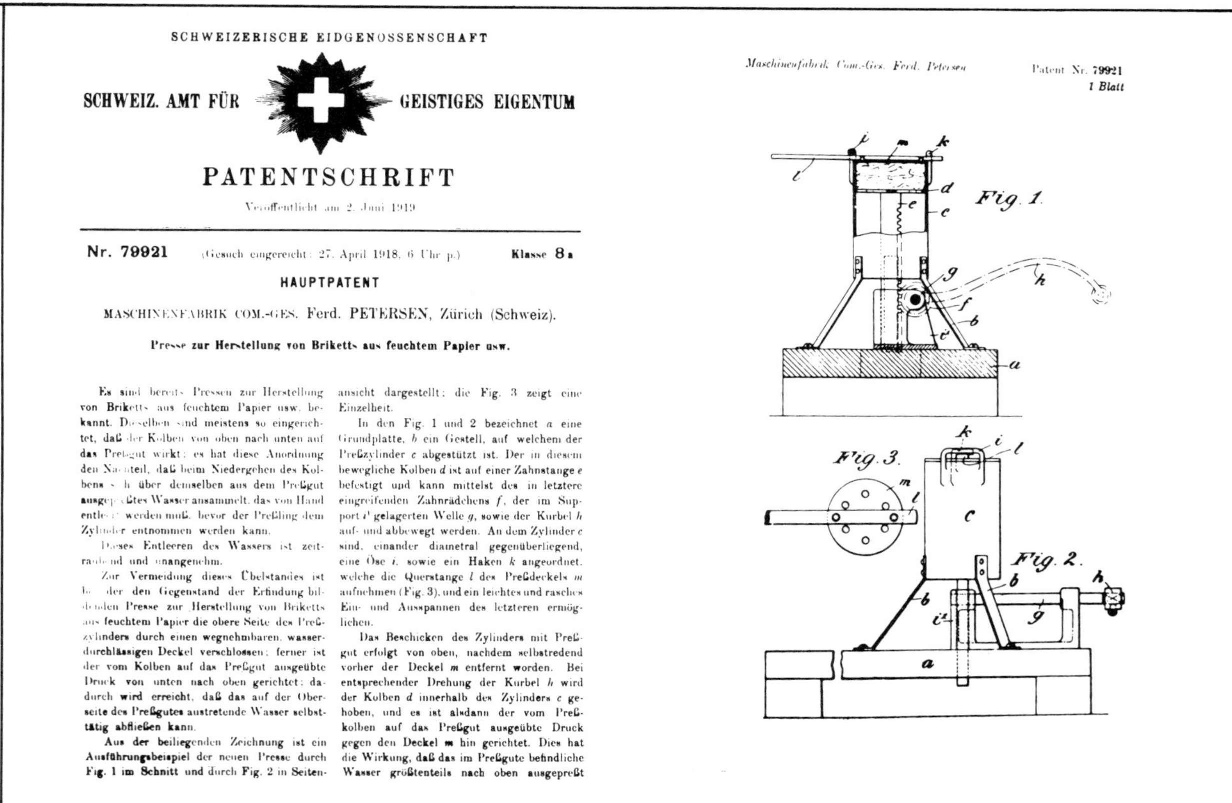

SCHWEIZERISCHE EIDGENOSSENSCHAFT

SCHWEIZ. AMT FÜR GEISTIGES EIGENTUM

PATENTSCHRIFT

Veröffentlicht am 2. Juni 1919

Nr. 79921 (Gesuch eingereicht: 27. April 1918, 6 Uhr p.) **Klasse 8 a**

HAUPTPATENT

MASCHINENFABRIK COM.-GES. Ferd. PETERSEN, Zürich (Schweiz).

Presse zur Herstellung von Briketts aus feuchtem Papier usw.

Es sind bereits Pressen zur Herstellung von Briketts aus feuchtem Papier usw. bekannt. Dieselben sind meistens so eingerichtet, daß der Kolben von oben nach unten auf das Preßgut wirkt; es hat diese Anordnung den Nachteil, daß beim Niedergehen des Kolbens sich über demselben aus dem Preßgut ausgepreßtes Wasser ansammelt, das von Hand entleert werden muß, bevor der Preßling dem Zylinder entnommen werden kann.

Dieses Entleeren des Wassers ist zeitraubend und unangenehm.

Zur Vermeidung dieses Übelstandes ist bei der den Gegenstand der Erfindung bildenden Presse zur Herstellung von Briketts aus feuchtem Papier die obere Seite des Preßzylinders durch einen wegnehmbaren, wasserdurchlässigen Deckel verschlossen; ferner ist der vom Kolben auf das Preßgut ausgeübte Druck von unten nach oben gerichtet; dadurch wird erreicht, daß das auf der Oberseite des Preßgutes austretende Wasser selbsttätig abfließen kann.

Aus der beiliegenden Zeichnung ist ein Ausführungsbeispiel der neuen Presse durch Fig. 1 im Schnitt und durch Fig. 2 in Seitenansicht dargestellt; die Fig. 3 zeigt eine Einzelheit.

In den Fig. 1 und 2 bezeichnet *a* eine Grundplatte, *b* ein Gestell, auf welchem der Preßzylinder *c* abgestützt ist. Der in diesem bewegliche Kolben *d* ist auf einer Zahnstange *e* befestigt und kann mittelst des in letztere eingreifenden Zahnrädchens *f*, der im Support *i'* gelagerten Welle *g*, sowie der Kurbel *h* auf- und abbewegt werden. An dem Zylinder *c* sind, einander diametral gegenüberliegend, eine Öse *i*, sowie ein Haken *k* angeordnet, welche die Querstange *l* des Preßdeckels *m* aufnehmen (Fig. 3), und ein leichtes und rasches Ein- und Ausspannen des letzteren ermöglichen.

Das Beschicken des Zylinders mit Preßgut erfolgt von oben, nachdem selbstredend vorher der Deckel *m* entfernt worden. Bei entsprechender Drehung der Kurbel *h* wird der Kolben *d* innerhalb des Zylinders *c* gehoben, und es ist alsdann der vom Preßkolben auf das Preßgut ausgeübte Druck gegen den Deckel *m* hin gerichtet. Dies hat die Wirkung, daß das im Preßgute befindliche Wasser größtenteils nach oben ausgepreßt

Maschinenfabrik Com.-Ges. Ferd. Petersen Patent Nr. 79921 1 Blatt

Figure 4.1. Patent specification and drawing of patented press developed for making logs from damp paper during World War I.

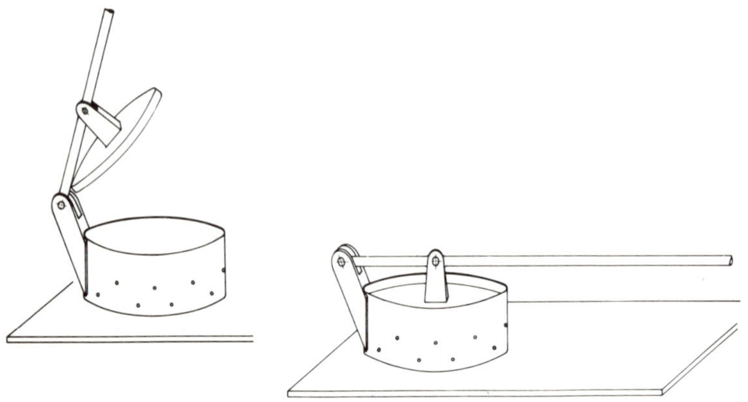

Figure 4.2. A simple household waste paper press used in Europe during World War II. Wet newsprint was used as raw material.

Significant increase in pressing power

For example, the old presses worked with a pressure of about 6,500 pounds per square inch (460 kilograms per square centimeter), while today we work with an average pressure of 17,100 pounds per square inch. On our most recent press of the FH 2-90-200 type, this means a pressing power of 152 tons on the two rams–which is more than 10 times the dead weight of the machine.

Such high pressure requires the very best material. For instance, bearing parts, although not visible, are extremely important. A technically-devised, thorough lubricating system, together with efficient oil filters, is imperative. But 152 tons of pressing power above all require the highest precision in construction of the machine.

I have mentioned the name Glomera. Formerly the Fred Hausmann presses were sold in the United States under that name. However, the name has been changed and all presses manufactured by my company now carry the Fred Hausmann name. Each of our presses is manufactured with the tolerances valid for machine tool construction. Our system is fully automatic, except for packing.

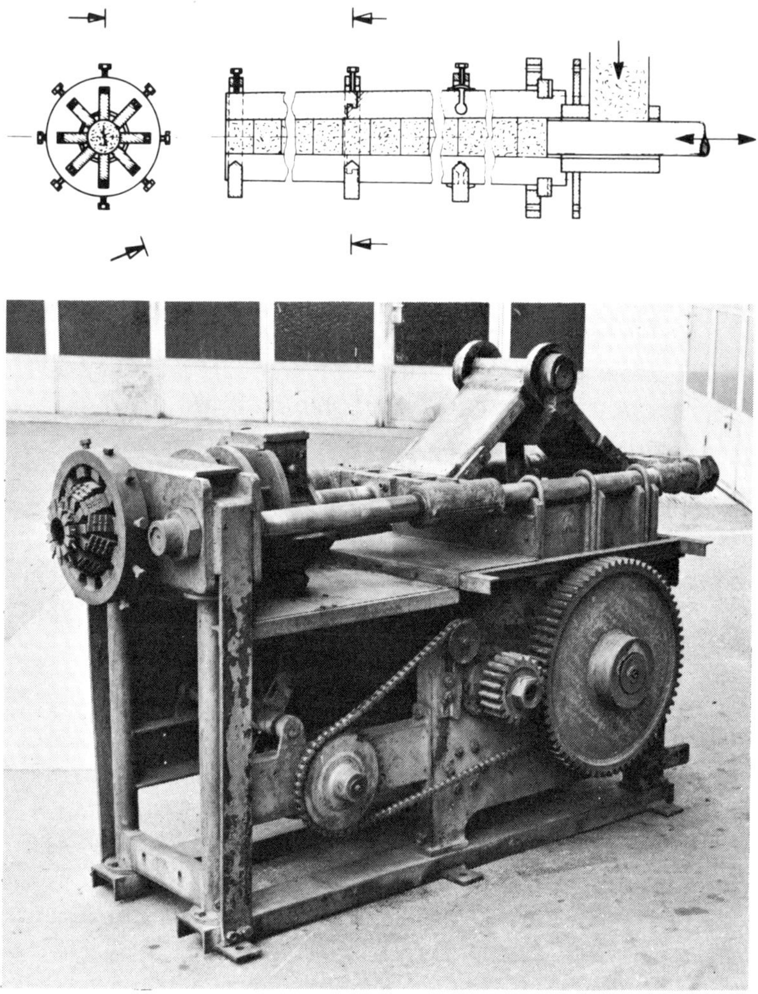

Figure 4.3 *Top*: Pressing dies patented by Caran d'Ache in 1942. *Bottom*: Swiss-designed cellulose waste press, 1942.

Briquettes solve European industrial fuel problems

A number of briquetting installations in Europe work without any attendants. Several firms store the wood waste briquetted during the summer months for use in winter to cover their energy requirements. Let me quote an example described in Great Britain's *Timber Trades Journal*:

> No-Nail Boxes, Ltd., subsidiary of Keizer Venesta, Ltd., Britain's largest importer and distributor of plywood, creates in its production processes tons of wood waste in the form of sawdust, small wood blocks, plywood offcuts, and trimmings. At first, the waste was burned in an incinerator. This not only contravened the Clean Air Act, but as Sir Gerald Nabarro, Chairman and Managing Director, said: "It is almost sacrilegious to waste the heat properties of refined wood products including all the resinous content and vast potential Btu's."

The Glomera briquettor was installed in the No-Nail boilerhouse as part of a trial scheme to heat the factory from its own waste wood products. Low-pressure steam from the boilers heats both factories–a total floor area of 180,000 square feet (about 18,000 square meters) at an even 65 degrees Fahrenheit through the winter. Summer output of briquettes is stored for winter use.

Financial savings realized

The financial savings are twofold: (1) no other fuel, such as oil or bituminous coal, has been purchased since the briquettor was installed; (2) the high cost of manually removing wood waste from the factories and disposing of it elsewhere is recouped–offering the major saving. The plant and equipment have an estimated life of 25 years, with reasonable maintenance. Sir Gerald Nabarro's own estimate is that, after taking as credit current investment allowances, the entire capital cost of the installation is recoverable in three years. After that period, the heating costs of the factory plus cost of waste collection represent an annual saving in fuel and labor of £5,000 ($11,750 approximately). "I have proven this process during the past seven years. I think it highly viable, economic, and a most attractive financial and fuel economy exercise," says Sir Gerald.

Another example is the Joka Furniture Mfg. Co. of Austria. By burning its briquettes, Joka accomplishes two things:

1. It is a "self-heating" factory.
2. It supplies the energy for the hot water heating system that kiln-dries the lumber used in manufacturing its products.

It is rare to get the best of both worlds, but Joka appears to be doing just that with its system that heats the plant and simultaneously disposes of all wood waste.

The Caran d'Ache Pencil Co., Geneva, Switzerland, one of Europe's largest manufacturers of pencils, has been recycling wood shavings, compressed in the Glomera briquettor, directly into the plant's furnaces. Here again the process is fully automatic and plant management expresses complete satisfaction, especially with the low-maintenance requirements.

Using the following illustrations, I would like to explain how a Fred Hausmann briquetting press works.

Briquetting press feeding device

Figure 4.4 shows the feeding device of the briquetting press; it feeds material to the ram and partially compacts it. At the same time it prevents backward expansion of the raw material during the forward stroke of the ram. This feeding device improves the rate of admission of the machine and with it, its output.

Principal elements of the briquetting press feeding device are (1) the housing; (2) the auger; (3) the auger drive.

Breakdown of wood elasticity is crucial

In a publication by Albert L. Stillman, which appeared in 1923, it was made evident that briquetting wood waste is possible only if one succeeds in breaking the elasticity of wood. The following is taken from Stillman's report:

> One vital requirement must be met in forming the briquette; namely, the elasticity of the wood must be destroyed either by high pressure or by previous heat treatment, or by a combination of the two; or the briquette formed will have no permanence. Excessive

pressures are expensive and can only be justified on something unique and valuable in the final product.

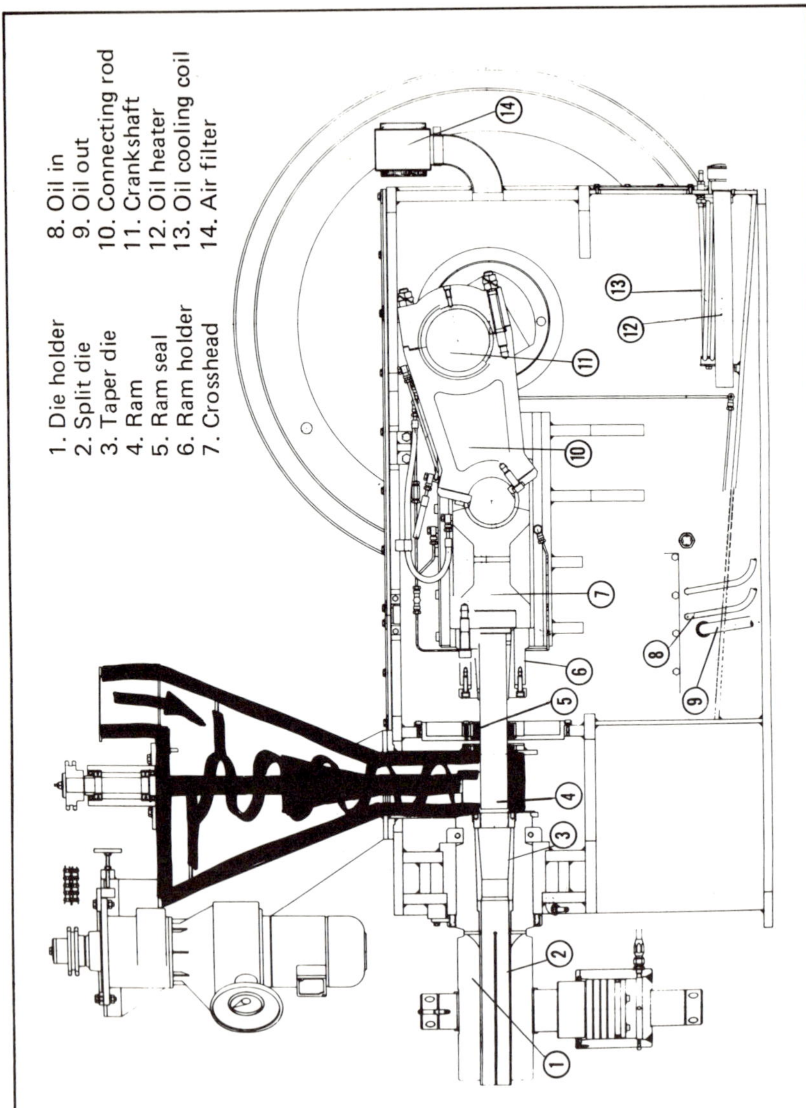

Figure 4.4. Feeding device of briquetting press (shaded area). Shape of the auger prevents backward expansion of raw material.

There is no agreement among experts whether the high pressure or the heating up of the raw material is responsible for breaking the elasticity of wood or other cellulose, thus making briquetting, or rather compressing, possible. Fred Hausmann presses use both high pressure and heat. The pressure exerted by our ram is 17,100 pounds per square inch. Depending on the raw material and, foremost, depending on intended use of the log, the ideal temperature will vary slightly.

Die holder the heart of the press

Heart of the briquetting press is without doubt the die holder. On the face, it appears to be a thick-walled pipe, but thorough study shows its rich technical inner life. Here is a short description of the die holder (shown in Figure 4.5):

1. *Seat and holding device of the interchangeable dies* (locking ring, taper die, split die). As the dies are subject to some wear and tear, they must be changed from time to time for new ones. All our dies are gauge manufactured. They will always fit the machine and easy replacement is guaranteed. The die holder is a patented part of the briquetting press.

2. *Hydraulic power transmission to split die.* The cylindrical split die has four longitudinal slots. By external hydraulic pressure, this cylindrical shape can be changed to a slightly *conic* form, to increase the friction of the log, thus increasing counterpressure as desired. The four slots of the split die continue outward in the four corresponding slots of the die holder. These slots are conic, to guarantee perfect operation of the hydraulic clamping device. The pressure on the die holder is adjustable from 5 to 35 tons (11,000 to 77,200 pounds-force).

3. *Die temperature.* During briquetting, the dies get hot. To avoid excessive heat, which would carbonize the log surface, the die holder is provided with a built-in cooling coil. This can be connected to the local water supply. The cooling water remains clean. It can be recirculated or connected to the sewer with no pollution problems.

Figures 4.6, 4.7, and 4.8 show the sequences of operation of three variants of briquetting installations: (1) for fine, dry materials; (2) for coarse, dry materials; (3) for coarse, wet materials.

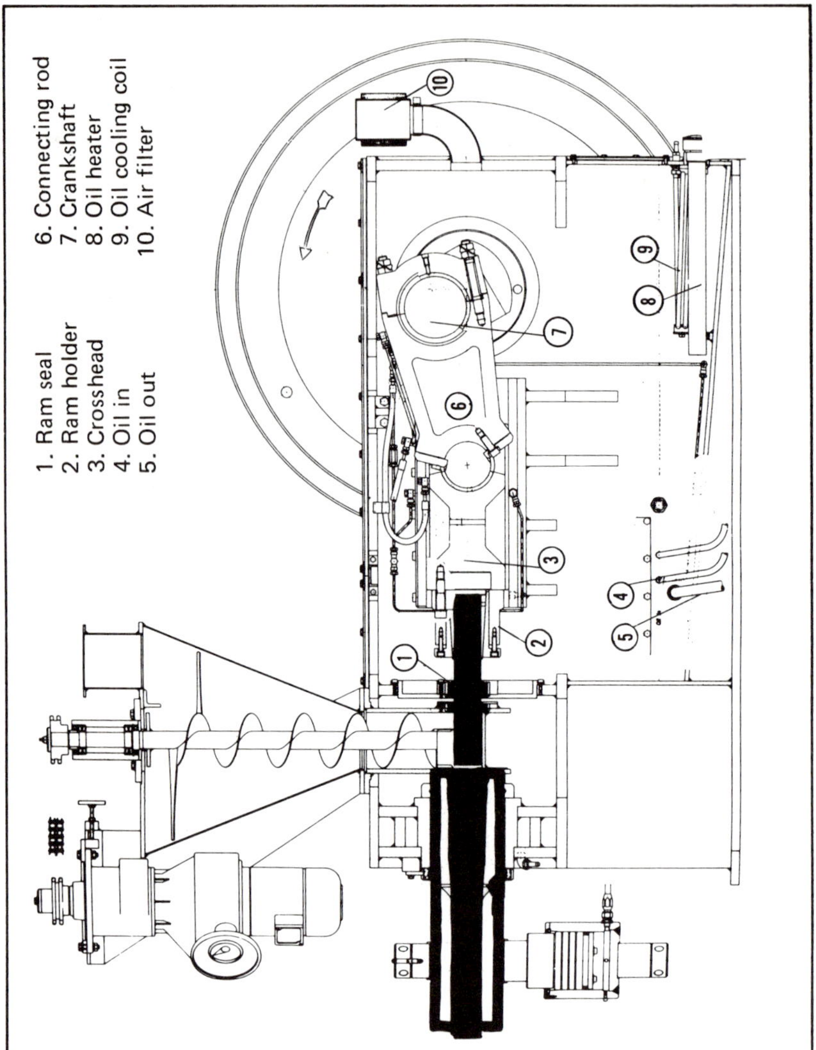

Figure 4.5. Briquettor for wood shavings. Shaded area is die holder, heart of Hausmann briquettor.

A simple briquetting installation

The simplest type of briquetting installation consists of a briquetting press and the feeding device. Figure 4.6 shows the feeding of a briquetting press with raw material from a silo through a vibrating chute.

The press also can be fed directly from a cyclone or by any conveying system (such as conveyor belt or screw conveyor). In any case, to obtain the best possible results, both in log quality and press output, the feeding of raw material to the briquetting press should be adjustable within certain limits. The type of briquetting installation shown in Figure 4.6 is suitable for briquetting material with a maximum moisture of 15 percent: dry sawdust, sander dust, shavings, peanut hulls or rice husks, etc.

An installation of this type with an output of 2,200 pounds (1,000 kilograms) of logs per hour consumes about 45 kilowatt-hours of power. An installation with an output of 6,600 pounds (3,000 kilograms) of logs per hour, consumes about 105 kilowatt-hours of electrical power during operation.

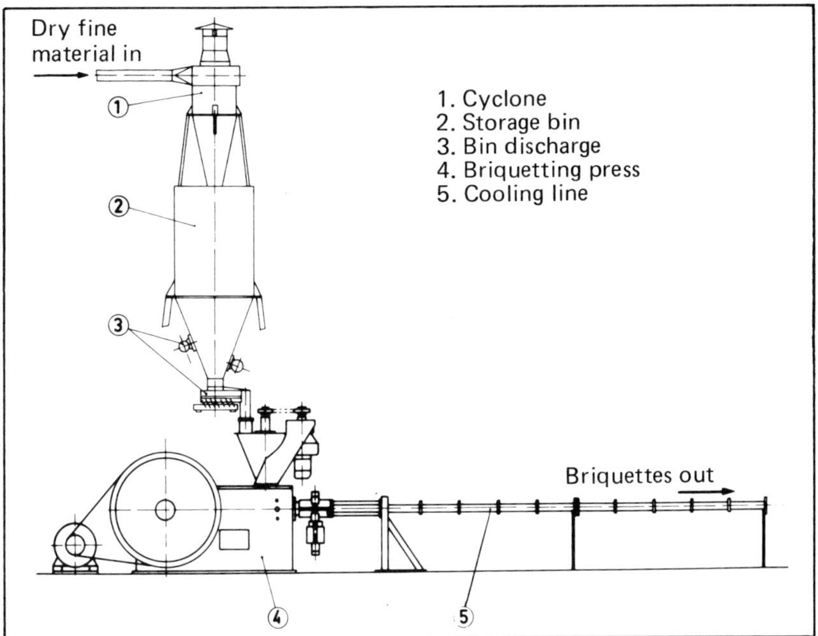

Figure 4.6. Dry fine material is fed from silo through vibrating chute into briquetting press (4). This is the simplest type briquetting installation currently in use.

Briquetting of coarse, dry materials

To briquette dry but coarse material, such as trimmings from a plywood or chipboard mill or from a furniture factory, the material must be ground prior to briquetting.

One approach is a double-rotor hammer mill, conveying device, bin, bin discharge, and briquettor (Figure 4.7).

Depending on type of material to be crushed, power consumption of an installation with an output of about 2,200 pounds per hour (1,000 kilograms per hour) is between 85 kilowatts when using a knife-type mill and 120 kilowatts when using a double-rotor hammer mill. Power consumption of this installation, with an output of 6,600 pounds per hour (3,000 kilograms per hour), is between 230 kilowatts and 270 kilowatts—also depending on type of mill.

Briquetting of coarse, wet material

The operating cycle of a briquetting plant for coarse, wet material is as follows. Wet raw material is hogged and conveyed to

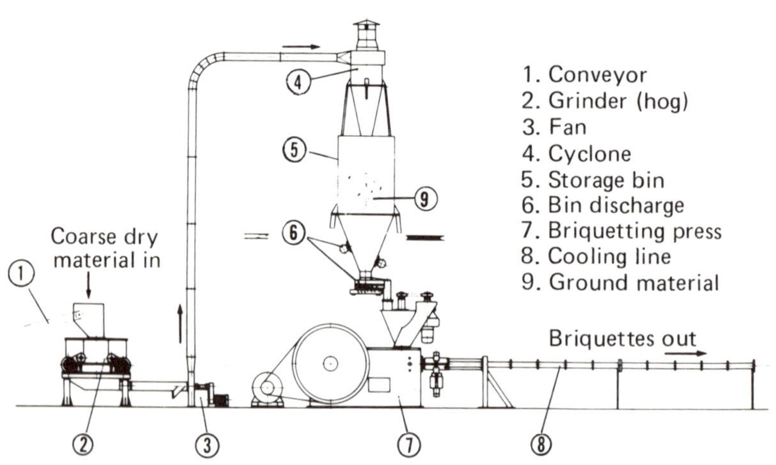

Figure 4.7. A cyclone-to-silo feed for handling coarse, dry material which must be ground prior to conversion into briquettes.

a silo. From the silo it is continuously fed into a current of hot air and conveyed to the dryer. A suspension dryer is used in the installation shown in Figure 4.8.

The dried material passes from the dryer to the cyclone and into the briquettor. The hot gas required for drying is produced by burning logs in a furnace. Any conventional furnace built for burning solid fuels can do the job. It is unimportant which type of fire grate the furnace has, or whether it is fed manually or mechanically.

Thermic drying is economically justified only up to 55 percent moisture of the raw material. With higher moisture, as for instance with wet debarking, we recommend that the material be predried mechanically. Ring roller presses are most suitable for predrying.

Processing engineering data for such an installation (without mechanical drying) with a log output capacity of 1 ton per hour, are the following:

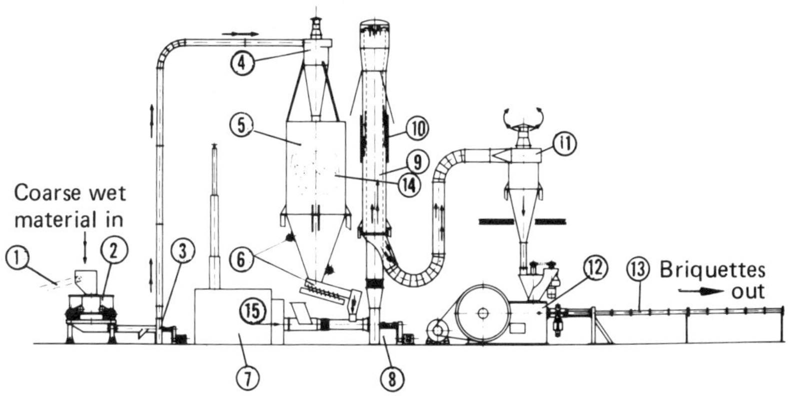

1. Conveyor
2. Grinder (hog)
3. Fan
4. Cyclone
5. Bin for wet material
6. Bin discharge
7. Furnace (briquette–heated)
8. Fan
9. Suspension dryer
10. Insulation
11. Cyclone
12. Briquetting press
13. Cooling line
14. Ground wet material
15. Hot air

Figure 4.8. Suspension dryer (9) on this installation enables briquetting of wet ground raw materials.

Feeding quantity: 3,970 pounds per hour (1,800 kilograms per hour).

Initial moisture: 50 percent of total weight.

Heat requirement for evaporation of water: about 4 million Btu per hour (1.0 gram calories per hour).

Log consumption to produce required heat: about 550 pounds (250 kilograms) per hour of logs.

Power consumption of installation: between 170 and 210 kilowatts.

The analogous data for an installation with a log output capacity of 3 tons per hour are:

Feeding quantity: about 11,900 pounds per hour (5,400 kilograms per hour).

Initial moisture: 50 percent water.

Heat requirement for evaporation of water: about 12 million Btu per hour (3 gram calories per hour).

Log consumption to produce the required heat: about 1,650 pounds (750 kilograms) logs per hour.

Power consumption: 420 to 480 kilowatts.

Firing and drying

The following explains the firing and drying installations.

A furnace for solid fuel has to meet two requirements. First, the fuel must be spread uniformly within the hearth. This can be done by hand (plane-grate type furnace); by chute (sloping-grate type furnace); or mechanically (spreader stoker, travelling-grate, or chain-grate stoker type furnace).

Second, the main combustion air must be distributed in such a way that it is fed uniformly or functionally to the fuel. This is accomplished by installing grate bars or grate plates, air gaps, zones, and also by feeding secondary air.

The dryer preferred with our briquetting plant is the suspension dryer (shown previously in Figure 4.8). It is characterized by simplicity: aside from a fan and an air lock, the dryer has no moving parts.

Suspension drying involves introducing small particles into a hot gas stream. The particles are conveyed and, at the same time, dried by direct contact with the gas. Gas temperatures are normally in the range of 500 to 1,500 degrees Fahrenheit. Contact time between the particles and the gas is short—a matter of seconds. Suspension drying is often called flash-drying because of this short contact time between particles and gas.

Successful drying of sawdust

It might seem that a heat-sensitive material, such as sawdust, could not be exposed to temperatures as high as 1,200 degrees Fahrenheit or more. However, upon entering the gas stream, the sawdust particle contains moisture and the particle tends to stay at the adiabatic saturation temperature of the gas, which is below 212 degrees Fahrenheit, until the moisture is evaporated. Heat for evaporating moisture must come from the gas, thus lowering the gas temperature. By the time moisture is removed from the particles, the gas is at such a low temperature that there is little tendency to char or burn the wood. Of course, particle size is important in performance of a suspension dryer.

In conclusion, let us return to John D. Rockefeller and consider what an enormous contribution you all can make to the solution of the energy crisis.

In the four Pacific states of Alaska, Washington, Oregon, and California, 21 million tons (dry weight) of unused bark and other wood waste are produced each year. (My source is the November 1973 issue of *Forest Industries.*) Twenty-one million tons of wood waste compressed into 21 million tons of Aglo logs equal 81 U.S. gallons of fuel oil per second (Table 4.1). Figure 4.9 shows the weight and heat value relationship of fuel oil and Aglo logs.

Conversion of that 21 million tons of unused wood waste into electric energy (Table 4.2) would provide 11 million kilowatts—or 19.5 times the generating capacity of the Bonneville Dam on the lower mainstream of the Columbia River.

In a speech early in 1974, the president of Shell Oil Co. discussed the extent of United States' dependency on foreign energy imports. He stated that if and when the Arab nations lifted the oil embargo, a daily gap of 260,000 gallons (1 million liters) of crude oil still would have to be covered.

Table 4.1. Comparison of Wood Waste Recoverable as Briquettes with Fuel Oil Consumption in the United States

21 million tons (dry weight) of unused bark and other wood residues are produced each year in the four Pacific states of Alaska, Washington, Oregon, and California (Source: *Forest Industries*, November 1973).

Compressed into 21 million tons of Aglo logs, these 21 million tons of wood waste equal the following amount of fuel oil:

$\frac{21,000,000}{365 \times 24}$	=	2,397 tons per hour
	=	4,794,000 pounds per hour
$\frac{4,794,000}{36,000}$	=	1,332 pounds of Aglo logs per second
7,830 Btu/lb	=	lower calorific value of Aglo logs made from average wood waste
17,640 Btu/lb	=	lower calorific value of average fuel oil
$\frac{17,640}{7,830}$	=	2.25 pounds of Aglo logs equal 1 pound of average fuel oil
$\frac{1,332}{2.25}$	=	592 pounds of average fuel oil per second
7.344 lbs/U.S. gal	=	specific weight of average fuel oil
$\frac{592}{7.344}$	=	81 U.S. gallons of fuel oil per second are wasted now

Eighty-one gallons per second equals 7 million gallons per day—which fills 27 times that daily gap.

The world reserves of raw and other materials are limited. The capability of our earth to absorb waste and pollution without danger is also limited. All over the world countries try to increase

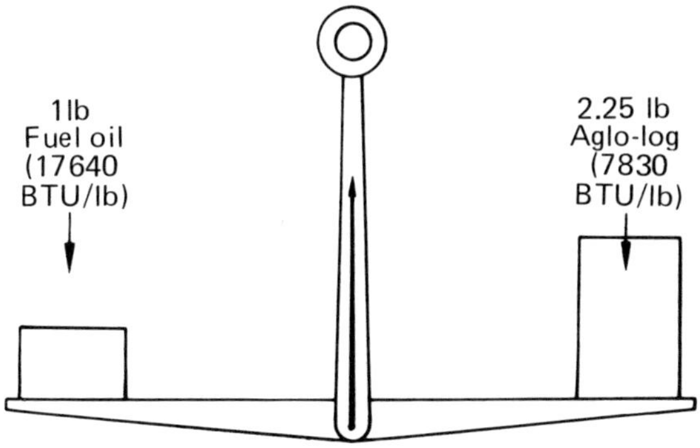

Figure 4.9. Relation of fuel oil and briquetted Aglo logs in terms of weight (pounds) and heat value (Btu's per pound).

Table 4.2. Potential Energy Supplied by Briquetted Wood Waste Compared with Generating Capacity of Bonneville Dam

21 million tons (dry weight) of unused bark and other wood residues are produced each year in the four Pacific states of Alaska, Washington, Oregon, and California (Source: *Forest Industries*, November 1973).

Compressed into 21 million tons of Aglo logs, these 21 million tons of wood waste will produce the following amount of energy:

$\frac{21,000,000}{365 \times 24}$	= 2,397 tons per hour
	= 4,794,000 lbs/hr
Lower calorific value of Aglo logs made from average wood waste	= 7,830 Btu/hr
4,794,000 lbs of Aglo logs per hour	= 37,537,000,000 Btu/hr = 37.5 billion Btu/hr
3,413 Btu/hr	= 1 kW
37.5 billion Btu/hr	= 11 million kW
Generating capacity of the Bonneville Dam (Columbia River)	= 564,000 kW
11 million kW	= 19.5 times this capacity

their national production, thus increasing their standard of living. Thus consumption of raw materials increases, again increasing waste.

Our civilization is based on the exploitation of natural sources of energy. Accordingly, a sure energy supply is one of the fundamental conditions of our economic life. Energy increases our comfort, the material well-being of mankind. We cannot produce more than the earth can give, and the environment can suffer, without danger.

That means: Don't dump it, squeeze it.

5

Bark breakdown for profitable products

Frank S. Trocino, Technical Director, Bohemia, Inc., Eugene, Oregon

Douglas fir bark has three physical components: cork, bast fibers, and amorphous powder. These parts are well defined in the bark. The cork of Douglas fir bark is more laminar in cell structure than is cork from the Mediterranean oak tree. The bast fibers are double-ended needle fibers about 15 times longer than their diameter. In addition to the third constituent—a fine, amorphous powder—wax is present in greater quantity than in most trees. Figure 5.1 illustrates the amount of bark available from old-growth Douglas firs.

Bohemia, Inc., has developed a solvent system to selectively extract the wax, most of which is in the cork. The process transforms the cork from laminar to cubical in structure, introducing many air pockets, which give a popcorn effect and resiliency to the cork. The remaining components of the bark are not physically affected by the solvent. After extraction, wax is separated from the solvent by simple distillation; the solvent is recovered and reused.

The recovered wax is a hard, light-green material classified as a vegetable wax. Most vegetable waxes, such as candelilla and carnauba, are imported. These waxes sell for between \$1.75 and \$2.75 per pound and are used in polishes, carbon paper, and

Figure 5.1. This photograph of an old-growth Douglas fir log indicates the very large portion of bark it contains.

cosmetics. After the wax has been extracted and the spent bark desolventized, many products can be made from the spent bark.

Plywood extender produced from bark

One use of the spent bark is to grind all constituents to pass through a 100-mesh screen. This product is called Bohemia extender and is used in plywood adhesives. At present the plywood industry uses predominately corn cobs, shipped from the Midwest and South, for extender in plywood adhesives. About 40,000 tons of extender—selling from $100 to $105 per ton delivered—are used each year in the Northwest and Canada.

Our solvent system makes mechanical separation of bark into cork, bast fibers, and amorphous powder relatively easy. The cork can either be sold as granulated cork or mixed with a binder and made into sheets. Sheet cork, 3/16-inch thick, sells for between 45 and 50 cents per square foot. Granulated cork, imported from Portugal, sells for about 50 cents per pound. Cork is made into bulletin boards, paneling, and gaskets.

Bast fibers give impact resistance to plastics. Competitive products are sisal and glass. Sisal is imported from Africa and sells for about 45 cents per pound.

Amorphous powder has been used as a phenol substitute. The basic ingredient in plywood adhesives is phenol, which is shipped into the Northwest and sold for approximately 25 cents to 30 cents per pound to adhesives manufacturers. In all cases, the products we have developed from bark are in short supply domestically and worldwide.

Bark supply compares favorably with other raw materials

About 5 million tons of bark are generated in Oregon and Washington each year. This figure does not mean much without some data for comparison. For example, the total supply of bark generated in the United States per year is 14 million tons. The total production of aluminum in the United States is 6 million tons, and we in the Pacific Northwest have a raw material volume almost equal to that. All our plastics—nylon, polyvinyl, and others—weigh in at 9 million tons per year. So—in bark—we have a raw material supply that far exceeds some of the biggest.

The light-colored portion of Douglas fir bark is cork. The brown streaks are amorphous powder, which has no cell structure. The third part is the prickly double-ended needles, bast fibers, with which you quickly become familiar when you handle Douglas fir bark—one reason Douglas fir bark is not popular on lawns, compared with hemlock and some other barks.

This pile of hogged fuel contains a high proportion of bark. About 5 million tons of softwood bark are generated yearly in Oregon and Washington.

Total log utilization

In bark preparation at Bohemia's Coburg, Oregon, plant, chunks from the debarker will be broken down to about 3-inch size. Material that will not be used in the extraction plant—sawdust, limbs, and so forth—will be diverted to the boiler to make steam for use in the extraction plant. Essentially what we have at Coburg is total utilization of the log. Logs coming into Coburg will make either veneer or dimension lumber and chips. The bark moves to a bark preparation and separation center. Part will go to the extraction plant; what is left will generate steam. We then will grind the bark into granulated sizes which will go to a dryer to be dried to between 5 percent and 10 percent moisture content.

Wax content high

What is unique in Douglas fir bark is its high wax content. All vegetable waxes used in the United States now are imported; one example is carnauba wax, which comes from the Brazilian carnauba tree. One use for vegetable wax is as an internal lubricant in all plastics. For instance, when polyvinyl chloride is made into a pipe, it is put through a die, involving much friction. So wax that is partially soluble in the polyvinyl chloride acts as a lubricant. The insoluble material stays on the surface and lubricates the dies. Millions and millions of pounds of wax are used in this way. In the manufacture of some plastics–polyethylene, for one–synthetic waxes are used. But almost all our synthetic waxes come from oil, and what is happening to oil is common knowledge.

Solvent key to wax extraction

We have developed a very selective solvent to extract the wax, which is located mostly in the cork portion of the bark. The solvent changes the cell structure of the cork from laminar to more cubical. This change opens up many air pockets–providing a popcorn effect–and gives a resiliency not previously attainable to cork from Douglas fir bark. We need to remember that the entire world's supply of cork comes from the Mediterranean oak tree, grown on tree farms in Spain, Portugal, and Morocco. It takes 150 years for the Mediterranean oak to mature. When harvesting the bark, the farmer strips only a certain section of the tree, so as not to kill it. And since trees can be stripped only once in eight years, stripping progresses from tree to tree. There was an attempt to transplant Mediterranean oak to California, and it thrived–but there was not enough bark for use.

The solvent selectively removes the wax, altering the cork only physically. But this alteration facilitates mechanical separation of the amorphous powder, bast fibers, and the cork. Then the solvent is removed. Figure 5.2 shows products that can be made from Douglas fir bark.

Wax produced from Douglas fir bark is very hard. We have made many alterations, chemically, to this wax. We have bleached it; we have used it in lipsticks. Ten thousand tons of wax–selling for \$1.00 to \$1.80 per pound–are used in the cosmetic trade alone. We also are trying to upgrade this wax physically so it can be used in carbon paper and as an internal lubricant in plastics. We have had good success. Demand for waxes now is tremendous. There

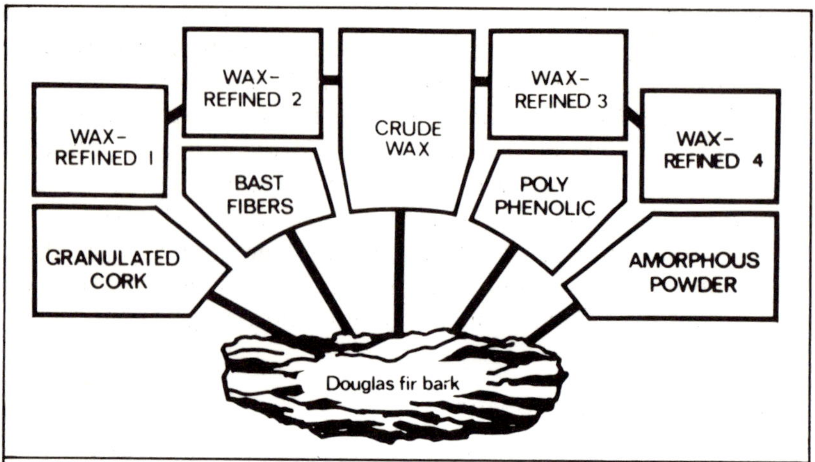

Figure 5.2. Many valuable products can be obtained from Douglas fir bark. Other softwood barks often lack its sliver-like bast fibers.

will be a shortage until at least 1975-1976. Since all wax is now imported, the ability to produce wax domestically will help our balance of trade. Wax extracted from Douglas fir bark will provide the first domestic source of vegetable wax in the United States.

After the extraction process, the greatest portion of the original raw material—the spent bark—remains. We desolventize it, recycling 99.9 percent of all the solvent, losing only 0.1 percent.

Standard equipment used

The equipment in the plant under construction at Coburg has been borrowed from the oil-bearing seed industry, which includes soybean oil and cottonseed oil. In the oil-bearing seed industry hexane is used as a solvent. We use a different solvent, but the only real difference in our work is that we are processing bark instead of oil-bearing seed. The equipment we will use is off-the-shelf. Hundreds of these oil-bearing seed extraction plants have been designed throughout the world.

Various products can be made from the spent bark. We can achieve complete mechanical separation by grinding it into an extender and putting it through a 100-mesh screen, as previously

Bohemia, Inc.'s fir bark plant under construction at Coburg, Oregon.

mentioned. Some 40,000 tons of corn cobs, mostly from Tennessee, are shipped into the Northwest to be used as extenders in plywood adhesives. We are offering a substitute—which has been approved by the American Plywood Association.

Why do we not simply grind the unprocessed Douglas fir bark, dry it, and use it as an extender? It cannot be done. It has been tried many times. The reason is this: wax is a fatty acid. Phenol-formaldehyde adhesives have a very high content of sodium hydroxide, which is a caustic. Years ago fatty acid caustic was used to make soap. When wax is added to a formaldehyde adhesive glue, it reduces the viscosity of the adhesive—which cannot be tolerated. Too thin an adhesive overpenetrates the veneer, resulting in delamination. An extender helps keep viscosity high. It is also cheaper than glue, so averaged in, it reduces glue cost.

Cork separation

We can separate the cork by a mechanical separation process, and a little know-how on grinding systems, to make the density of the cork as low as possible. Oak cork shipped from Portugal and sold here in bulk has a density of between 3 and 3½ pounds per cubic foot. Initially, the bulk density we were able to achieve with

Here is the full Bohemia complex during erection. Much of the bark to be processed at this installation will come from the sawmill at rear.

Douglas fir bark was around 6 to 7 pounds per cubic foot, but we finally achieved a breakthrough. We became a little more sophisticated in our pilot plant in Eugene, Oregon. (It is just a miniature plant. We hope to make our mistakes there and gain a little sophistication in choosing equipment.) Now we are producing cork with a density of 3 to 3½ pounds per cubic foot. We can grind this for sale as granulated cork to companies that make gaskets, bulletin boards, and similar items. Gaskets are the biggest outlet for cork. We have made some very successful pilot runs with a company making gasket material, and have achieved the necessary 85 percent resiliency. So we may adapt our equipment to make this type cork. We have a small press with which we have made this cork into sheets successfully.

Phenol substitute possible

Now let us consider amorphous powder. Amorphous powder contains phenolic groups. These groups are not unlike the phenol used in phenol-formaldehyde adhesives. Phenol sells for about 25.5 cents a pound, and I believe about 500 billion pounds of phenol-formaldehyde are used in the plywood industry. We have been able to substitute amorphous powder for a certain portion of the phenol that is derived from oil-based industries. Our substitute is not a mere extender. We are working for approval by the American Plywood Association of amorphous powder as a phenol substitute in the manufacture of adhesives.

Bast fibers for the plastics industry

Bast fibers give impact resistance to various plastics. Pure plastics shatter under a blow. So manufacturers add all kinds of fillers and fibers, such as fiber glass and sisal imported from Africa. We have developed a filler which can give impact resistance to most plastics.

Why all these products? In our research into bark utilization, we have tried to use the philosophy that is basic to the oil industry. Crude oil goes into many products. The product depends on the characteristics of the crude oil, since physical characteristics of oil vary. But the oil industry produces many products—distillates, gasoline, jet fuel, lubricating oils, and so forth. This is our philosophy with Douglas fir bark. We know that there is a difference between bark from old and young growth Douglas fir. We plan to adapt our processing as the raw material changes.

DISCUSSION PERIOD

QUESTION: Does the cork expand?

TROCINO: That is right. The solvent system expands the cork from its original size, because it develops air pockets in the cell structure.

QUESTION: Will these bark products compete on the open market?

TROCINO: Yes. The wax competes with wax that costs $1.35 to $2.75 a pound. The extender competes with corn cobs coming in for $100 a ton, or 5 cents a pound. Bast fibers should draw from 35 to 50 cents a pound, competing with sisal and glass. Amorphous powder competes with a phenol substitute that now costs 27.5 cents a pound. So we are talking about competition with products that are high in price.

QUESTION: Do you intend to use bark from other species?

TROCINO: Right now we are using only Douglas fir bark. All plant material contains wax, and these waxes vary. Some are very hard; others are very soft. Douglas fir has a hard wax that we can use. But eventually I think we will be able to use wax from other species. Waxes, being fatty acids, vary in length like a chain. The molecule is like a link. You can vary this chain by varying its length, thereby changing the physical characteristics of the fatty acid. Those fatty acids with the right chains can be used as lubricants–for example, in automobile automatic transmissions. (This is a lubricant which has been very scarce.) So there are many possibilities with further research.

Remember that until the late nineteenth century the chemical industry was based on destructive distillation of hardwood. All our chemicals came from wood. Methanol–which we call wood alcohol today–is such a synthetic. Acetic acid, formaldehyde, ethyl acetate–all these chemicals were derived from wood at one time. In the past chemical producers went into the hardwood forests, harvested the wood, brought it in, and destructively distilled it. The entire process was quite expensive. Then someone in Pennsylvania, instead of getting water from a hole, got "rock oil." When that oil was discovered, it really shot the whole wood chemical industry. Now we are going back to the forests; but today we are a little more sophisticated. Now the expense of harvesting the log is shared by many products. So it is in the oil industry. But fossil fuels are becoming scarce; we must look to chemicals that we can get and use. The cycle is repeating itself. Eventually, I think, we are going to see more and more chemicals derived from wood. So the whole picture changes. Glues change. In the next 10 years—once we become more sophisticated in determining what these reactive groups are in wood—we will be utilizing these reactive groups to make adhesives for use in the plywood and particleboard industries. I feel this is coming.

QUESTION: Are you having problems with pollution controls?

TROCINO: No, not really. The only place we are having a few problems is in our boiler. However, we think we can overcome that. To make this operation economically feasible, we must achieve very high recovery efficiency in the extraction process. This has already been proven in oil-bearing seeds; in California, extractors of cottonseed oil are getting about 99.9 percent recovery of solvent. This solvent does not differ greatly from the solvent in paints—except there is far more pollution from painting than there ever will be from this plant. There will be no odor. If there is an odor, it means we are losing solvent and we had better find the leak.

QUESTION: Are you pressure extracting? And what is your solvent?

TROCINO: No, we are not pressure extracting. We extract at atmospheric pressure. The solvent has been patented by Bohemia. We call it "mixed aliphatic aromatic solvent."

QUESTION: How did you decide what size plant to build?

TROCINO: First, we realized that we had to sell the spent bark. The extender is probably the cheapest return we can get. There is a 40,000-ton-per-year extender business. We postulated that we would try to capture half that business. So that is the way we designed our plant. That takes 2,000 tons of dried bark per month and will solve our bark problem at Coburg. We are hoping, with the other products, that we can build a plant at each site so we have total utilization. Remember too, we do burn. We still need fuel.

Today, the wood industry is sitting on top of the world, for one reason alone–it is the only industry that can satisfy its own energy problems. None of the others can do it; they must have oil, or something. But we in the wood industry can take what we cannot turn into products and burn it to make energy for drying and for process steam in other plants.

QUESTION: Can this process be made available for use by other companies?

TROCINO: This is for upper management to decide.

6

Computerized planning for maximum profit

Dr. Eugene L. Bryan, President, Decision Dynamics, Inc., Portland, Oregon

John R. Lang, Manager—Planning & Administration, U.S. Wood Products Division, Evans Products Company, Portland, Oregon

Part one by Dr. Eugene L. Bryan

A sawmill manager has three major areas of concern: his raw material, his production facilities, and the markets for his products. The manager's task is to coordinate all the variables relating to each of these areas in such a way that he gains the *best possible* return on his company's investments.

In sawmilling, the variables are so numerous and their interactions so complex that the *best possible* performance cannot be accomplished without the use of computers. This discussion will examine the complexity of management decisions and describe the application of computerized procedures for establishing optimum operational activities and policies.

Decision areas

A manager has a number of environments that influence his decisions: his own internal environment; an industrial environment representing the state of the forest products industry and its markets at any point in time; the U.S. economic and political environment; and a world trade environment. Of these, he can control directly only his own internal environment. As outside conditions continually change, the manager must shift his operation to stay in accord with those external factors that influence his company's welfare.

This requires timely information and timely action. Specifically, the manager has to know what he should be doing at any point in time to maintain the best possible performance for his operation. He must know what he should be doing at any point in time with his raw material options, his production facilities, and his market opportunities. The computer can be a very effective tool in helping a manager keep his organization operating in full context with the "outside world."

The use of computers to aid in management decisions requires a very well defined purpose and continued focus on that purpose. A manager's purpose, whether he is in sawmilling or any other industry, is *to achieve the maximum return on his investments through the best assignment of resources under prevailing conditions.* It is frequently easy to lose sight of this purpose in the confusion and disruptions of day-to-day business activities.

Computerized programs have been developed that can be aimed specifically at providing the information a manager requires to achieve maximum profit levels. When properly employed, these programs become tools which guarantee continued focus upon the primary purpose of a company's existence and which make possible the full achievement of that purpose.

Linear programming

One computer-based aid for decision making involves a technique known as linear programming. This involves a procedure designed to establish the *optimum allocation of resources* for a specified purpose—such as, maximization of profit.

A definition of terms will help in understanding the use and value of linear programming. Here, the word *optimum* means the best under prevailing circumstances.

Allocation means an assignment when faced with alternatives. Here the key word is *alternatives*—if you do not have alternatives, there is no need for computerized procedures.

Finally, the term *resource* is defined broadly as that which can be used to achieve a purpose. Under this definition, resources involve such things as capital, trees, logs, labor, machine time, inventory space, sales capabilities, and management time.

Essentially, in directing the way to the optimum allocation of resources, linear programming can be used to show how available resources should be used at any point in time to achieve maximum return on a company's total investments.

Optimizing resource use

Optimum allocation of resources is quite a complex problem. Each resource is a variable which can be controlled to some degree. Each variable interacts with other variables to provide a very large number of alternative operating patterns. This is illustrated in Table 6.1.

The table offers a simplified example which clearly illustrates the complexity of management in the forest products industry. In this example there are: ten ways to invest available capital; five sources of timber with three alternative bucking policies for each source; five log sizes (diameter groups) with three grades and three possible uses for each log type—they could go to a plywood plant, a sawmill, or the open log market; twelve labor alternatives involv-

Table 6.1. Possible Activity Patterns Multiply

Resource		Number of alternatives
Capital	[5 possible investment programs] x [2 forecasts]	= 10
Trees	[5 sources] x [3 bucking policies]	= 15
Logs	[5 sizes] x [3 grades] x [3 uses]	= 45
Labor	[6 departments] x [2 manning policies]	= 12
Machines	[6 work centers] x [3 options]	= 18
Space	[3 areas] x [3 inventory policies]	= 9
Sales	[2 markets] x [3 choices of emphasis]	= 6

10 x 15 x 45 x 12 x 18 x 9 x 6 = 78,732,000 choices

ing manning schedules; six work centers, each with three options available on the use of the machines; three areas of inventory (for logs, work-in-process, and finished goods) each with three alternative policies; three choices of sales emphasis for two different markets. The mathematical combinations of these resource alternatives number in excess of 78 million choices of activity that a manager would have to face if he were to consider all alternatives available to him.

Juggling the variables

At peak performance, Albert Einstein could deal with no more than six variables at a time before bogging down in the confusion of inevitable interactions. As seen in the above example, forest products managers face hundreds of variables which multiply into millions of alternative operating patterns. These variables can be dealt with a few at a time through conventional procedures. But without the use of a computer, it is virtually impossible to be aware of all alternative patterns and to choose the best combination of activities. By programming all of the variables that affect profits, the computer can help a manager consider all the millions of possibilities he faces to determine the best single choice of activities for maximum profit.

What is required to deal with the complex environment of a lumber operation—and to deal with it effectively and most profitably—is a powerful tool. Linear programming is such a tool. It will allow a manager to consider all his resources with their various alternatives, all production variables, all market opportunities, and all interactions of all variables related to these areas of concern.

Linear programming uses mathematical models to represent real world activities and to weigh alternative courses of action. It is important to understand just what these models are. Physical models are used in many professions to safely test new ideas, to measure the effects of various forces, and/or to observe results of alternative procedures or policies before full-scale commitments are made. For example, an engineer would not consider building a bridge before testing his ideas and calculations through modeling.

In general, models provide sources of essential information on a proving ground of minimal cost and consequence.

Testing decisions

In the past, business managers often had no choice but to test their most important and complex decisions on a full-scale basis in the real world—with severe consequences for errors and omissions.

The advent of management science and the powerful computers of today have made available models for managers. Since a properly constructed model accurately simulates all important aspects of an operation, any contemplated changes in log supply, machinery, labor usage, and/or market emphasis can first be tested through the model to determine *in advance* the effect of such changes on all aspects of the operation, including their effects on profits.

Following is a list of typical information that can be made available through use of models and modern management procedures:

- Optimum log allocation between alternative uses.
- Log breakdown patterns for maximum profit.
- Optimal inventory policies.
- Dollar values of each log type for each alternative use.
- Breakeven levels for bidding on specific timber sales.
- Optimal bucking policies.
- Optimum product mix under current or forecasted conditions.
- Work center assignments for optimum product mix and related production scheduling information.
- Quantitative information for sales planning.
- Evaluations of new equipment proposals.
- Effects of any contemplated change in log supply, plant facilities, or sales policies.
- Objective answers to any specific questions relating to modeled operations.

The use of computers in management decisions is not just a theoretical game people play for academic interest. It is a viable, practical, highly profitable technique and it is being used right now by a number of different companies. One example of a highly successful application of a linear programming model will be described in part two of this chapter.

Part two by John R. Lang

Part two of this chapter will describe how the Canadian Division of Evans Products Co. was able to increase the profits of a dimension sawmill without buying excessive new sawmill equipment. Our approach was to use a three-man task force of company employees plus assistance from Decision Dynamics. The results are not theoretical. They are results that we obtained with the people whom we employed at the beginning.

From the planning standpoint, all key variables and decision points were put together in a single computer model. Consequently, each variable–i.e., log cost; recovery by sawing pattern–can be considered simultaneously with all other variables. Changing one condition (variable) automatically affects the entire model. Each new printout from the model shows the most profitable operating policy for each major work center.

Our sawmill is located in the southern interior of British Columbia, just west of Banff National Park. Timber is lodgepole pine with spruce, fir, hemlock, and some cedar. All logs with a 4-inch or larger top must be brought to the log yard. The logs are scaled by volume with a "cunit" measure–100 cubic feet of log.

Accurate diameter estimate vital with small logs

I would like to point out specific places we found to be critical to sawmill profits.

Figures 6.1 and 6.2, respectively, show the log input and lumber output from a chipping headrig operation. The operator must determine the diameter of each log and select a setting (from among 10) that will give the best possible lumber recovery. Obviously, we can all get chips out of a tree. But the trick is to get the most lumber out of each log while maintaining production in excess of 2,000 logs per shift.

We asked ourselves, "How critical is this man's judgment in determining the effective log diameter at the chipping headrig?" For example, in Figure 6.3, the log in the top diagram is only 0.6 inch in diameter less than the log in the bottom diagram, yet the top log yields only 2x6 and 2x4, whereas the bottom log yields 2x8 and 2x4. That is a significant difference in terms of lumber

Figures 6.1 and 6.2. Log input and lumber output of typical Chip-N-Saw run at Evans Products Co.'s mill in British Columbia southern interior.

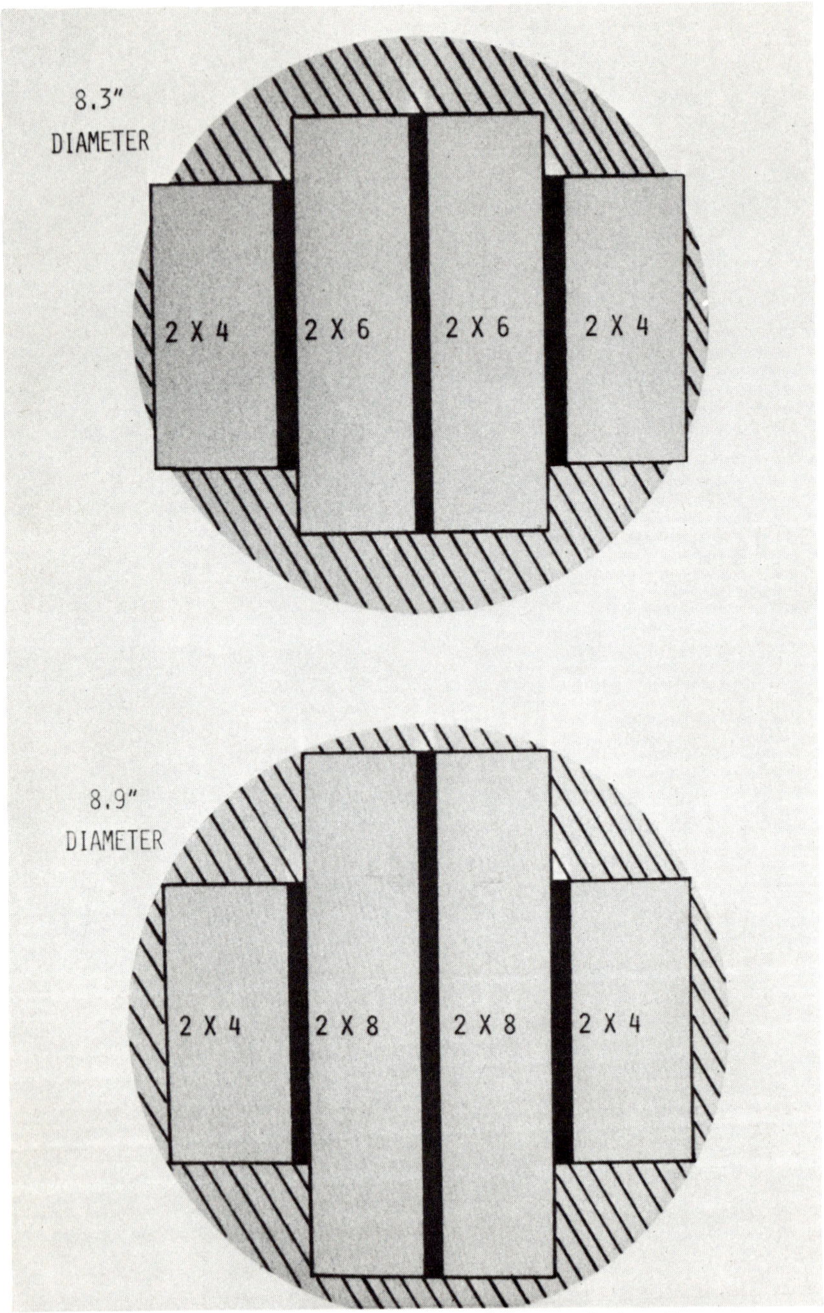

Figure 6.3. Jump in yield from very small difference in log diameter.

recovery. However, 0.6 inch is not discernible to the human eye.

We decided to install a minicomputer with a log scanner to help the operator improve his lumber recovery. There is a table of saw settings stored in the computer for each log length and diameter class. Having scanned the log and determined its length and effective diameter, the computer selects the proper saw setting from the table. The operator feeds the log to the saw and the computer begins to scan the next log. This machine is truly a production control device.

It has, as a by-product, quite a range of accounting and production data. It records the number of pieces put through the Chip-N-Saw each shift. In fact, we found that the fellows were not getting 2,000 logs a shift. They were just a little heavy on the counter button. Secondly, it records very accurately (in 2-foot increments of dimension lumber) the length of the sawlogs coming into the mill. Further, it gives us an accurate count of logs by diameter class. We are very concerned with allocating each log to the best facility since we have a chipping headrig, a bandmill, and a plywood plant operating less than 15 miles apart.

With the minicomputer we also get accurate measurements of the volume of logs having gone through the sawmill. Consequently, we now have a better measure of lumber recovery. The number of times that the operator overrides the computer's decision is a direct measure of the effectiveness of the entire program. Our experience to date has shown the number of overrides to be less than 10 percent of the logs processed.

Effect of woods bucking policy at bandmill

Moving from the chipping headrig mill to the bandmill, we used dollar value per cunit of sawlog to evaluate our woods bucking policy (Figure 6.4). The vertical axis is the dollar value per cunit. The horizontal axis is increasing log length coming from the woods. One can see that the dollar value per cunit increases as log length increases. This is to be expected considering the length premiums for dimension lumber. However, values begin to level out and suddenly drop at mid-range. We found that the longest log is not necessarily the most profitable log.

Having analyzed woods-length logs, we should review operating

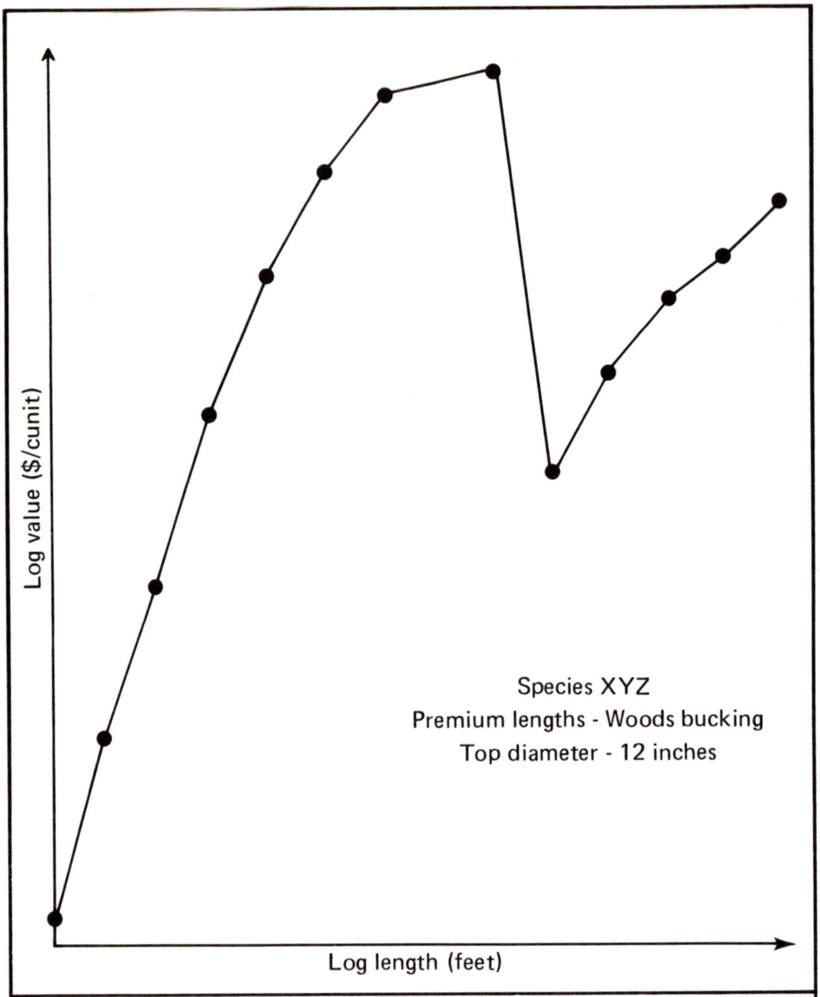

Figure 6.4. Consideration used in determining a woods bucking policy.

instructions at the bucking station. The key data at this station concerns log diameter, length, and grade. Figure 6.5 shows a 48-foot log with four of the bucking alternatives that we put into our model. The top diagram illustrates a traditional sawmill approach of bucking the log into three 16-foot logs. The second combination produces three logs also (18, 18, 12). (In certain diameter classes, there are tremendous premiums for each of those particular lengths.) The third combination involves only two logs and requires a full 24-foot sawmill. (Two long logs will probably increase production as compared with three shorter logs. However, this alternative requires close attention to taper.) For the fourth combination, we assumed the entire 48-foot log to be of peeler quality. At the large end, we have taken four peeler blocks, leaving a 14-foot sawlog.

The computer program analyzes all four of those bucking combinations (plus others) for each log species, grade, and diameter class. The "best" answer depends on all of these characteristics, plus production and market conditions. Of course, the best answer relates to profit. The program selects the most profitable alternative and prints the answer in a form that can be translated into rules for the bucking operator. His judgment is still an important part of his job. But our purpose is constantly to remind him of end product values.

Choice of breakdown pattern and sawing method

As the logs move into the bandmill, we should be looking for additional decision points. Our mill has two headrigs that are used

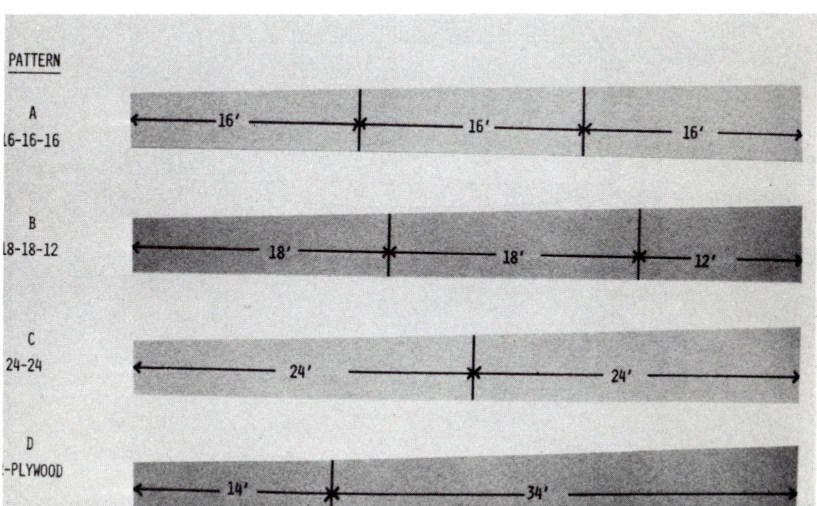

Figure 6.5. Alternative bucking programs for a 48-foot log.

for the initial log breakdown. Play sawyer and think about what you are going to do with that next log. Assume it is 12 inches in diameter. One of many breakdown patterns (for dimension lumber) that can be achieved with a 12-inch log is reflected in Figure 6.6. It happens to be live-sawing with the emphasis on a 12-inch program. In other words, we are looking for 2x12. One could redraw this diagram to emphasize 10-inch or 8-inch or to emphasize cutting timbers out of that sound wood. We did our homework—we drew the circles and the results changed our entire breakdown policy.

Deciding what to do at the headrig forced us to think about the various log diameters that we might receive there. Should both large- and small-diameter logs be live-sawn at the headrig? How should we use the quad and the vertical resaw? By emphasizing the loading of the machinery—the mill flow—as well as recovery, we were able to determine the best breakdown pattern for each log type received at the mill.

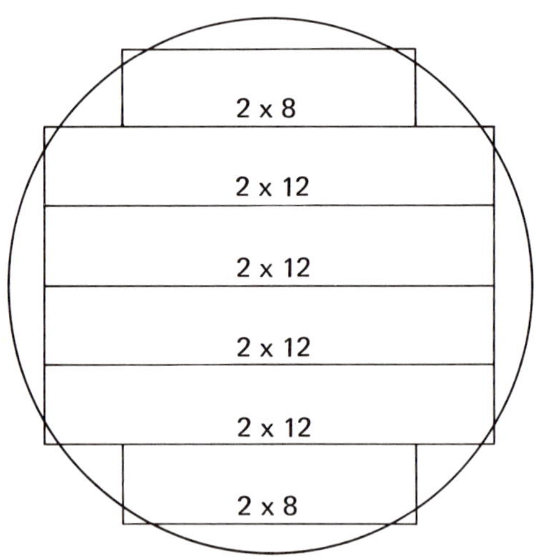

Figure 6.6. Possible breakdown pattern for a 12-inch log, live-sawing.

Decisions in sorting and the question of grade

The last major decision area in our bandmill is the green chain. How should this lumber be sorted? We added a lumber grader to improve the way we accomplished this. In other words, we believe that lumber grade, as well as length and width, should be considered at the green chain. Figure 6.7 illustrates green chain instructions for our dimension lumber by size and grade. The computer was used to make these decisions at the green chain by looking ahead to the ultimate destination of the lumber.

	TO DRY KILN			GREEN SALES		
	COMMON	STANDARD BETTER	UTILITY ECONOMY	COMMON	STANDARD BETTER	UTILITY ECONOMY
2 X 4				X		
2 X 6	X					
2 X 8		X				X
2 X 10	X					
2 X 12		X				X
BOARDS				X		

Figure 6.7. Instructions for sorting by size and grade at green chain.

Notice the distinction between the lumber sent to the dry kiln as compared with the lumber we plan to sell green. Within each size category, the report shows what to do with lumber between green chain and boxcar.

Another sawmill decision point is the tradeoff between width premiums and grade recovery in dimension lumber. What would be the possibility of improving grade by splitting a low-grade 2x12 either into a pair of 2x6s or into three 2x4s? Recall that we segregated our low-grade 2x12 at the green chain. Consequently, this 2x12 would have been pulled into a load containing both Utility and Economy lumber.

In Figures 6.8 and 6.9, the dark area represents a knot or other defect. The dotted lines indicate where we might split the piece. With the planer running at capacity, there is no time to orient the pieces. Test runs on this type of lumber will provide the grade recovery data necessary to answer this question. The market value

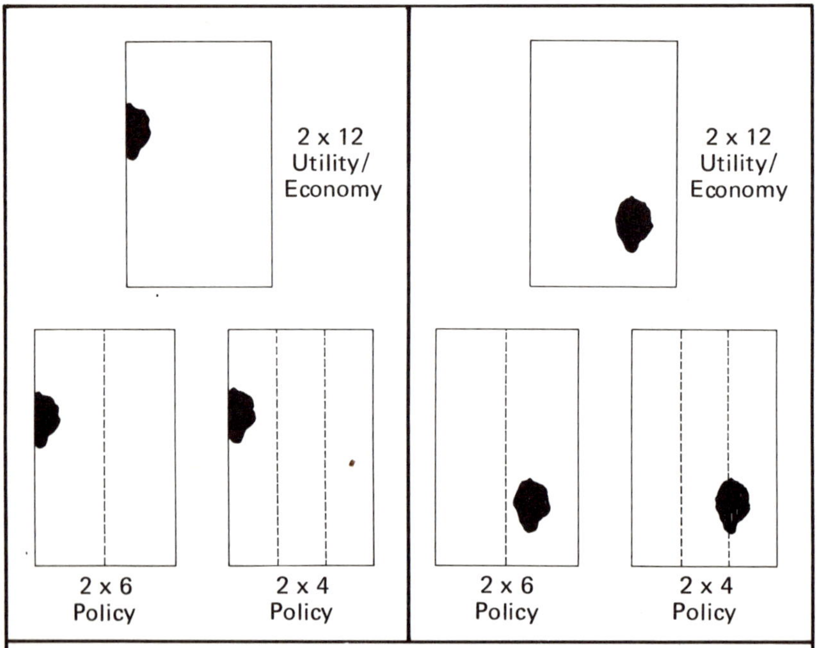

Figures 6.8 and 6.9. Some alternative ripping patterns for defective 2x12 dimension lumber.

of this grade improvement (if any) will be calculated by the computer model based on current market conditions.

Conclusions

We found that straightforward decisions such as the ones described above could be modelled by a computer program. We can enter a set of log deck characteristics along with current lumber prices, and the computer will determine what to do at each major work center.

Now that the sawmill model is operating, we have begun to explore other alternatives. Figure 6.10 shows individual log values (in dollars per cunit) based on the log's use as veneer, plywood, or lumber. In this example, the highest dollar value per cunit of log would be in plywood. (The plywood and veneer curves are fairly smooth, whereas the sawmill curve produces point-to-point situations because of the predetermined diameter groups used in our sawmill data collection.)

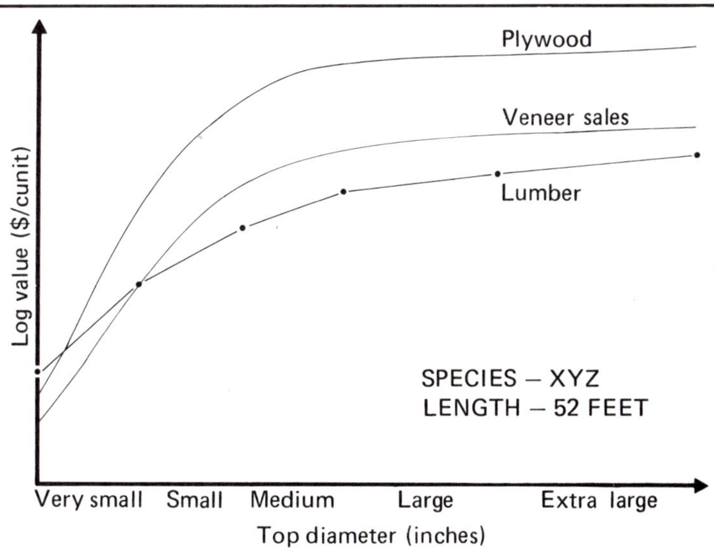

Figure 6.10. Comparative values for the log if manufactured into veneer, plywood, or into lumber.

It is very important to us to know when a log is more valuable as lumber than as plywood or veneer. To help make this decision, we can generate a log allocation plan by putting models together—i.e., a plywood model and a sawmill model. The models are small enough to be run on our computer and we maintain the data ourselves. So we are self-sufficient.

At first, we got outside help from Decision Dynamics. We are happy for that. On the other hand, the analysts in our Operations Research Department now maintain this model and are building other models for other species and other sawmills.

DISCUSSION PERIOD

QUESTION: What is the average length of logs run through your Chip-N-Saw?

LANG: Average length is in the 15-foot range.

QUESTION: What is the cost of a project like this in dollars per thousand feet over a year?

LANG: I do not think it is fair to state it in dollars per thousand in our particular case. For the money we spent on the mini-computer plus outside consultants, we anticipate a payback in less than one year. Single mill operators could get their money back within a year if they produced at least 25 million board feet per year. Of course, we have several mills. In that case, the cost per thousand and the out-of-pocket cost per project drops dramatically.

QUESTION: As you were describing results, you listed some general areas of improvement. I do not know whether you care to mention that here, but I think it would help to give a feel for whether or not this is going to be worthwhile to explore.

LANG: We used our own in-house computer, a linear programming package, and the services of Decision Dynamics to achieve dramatic results at this mill. These results specifically, as related by the mill manager, are: increased lumber recovery, increased grade recovery, increased sales realization, and increased sawmill productivity. In other words, these four fairly well recognized barometers all point in the right direction. Profits went up accordingly.

7

Orienting wood fibers for added strength

Ron Dorsett, President, MK-RDA, Inc., Portland, Oregon

Wood is a grossly misunderstood product of nature. Even within the wood industry we have a tendency to believe that we have gone about as far as we can go. However, results from the Washington State University (WSU) laboratories at Pullman, Washington, have demonstrated that we are, in fact, on the threshold of a whole new concept in the conversion of wood to lumber and other products. WSU has awarded my company, MK-RDA, Inc., the opportunity to commercialize their research. Our research activities involve taking basic conclusions arrived at in the WSU laboratories and developing the necessary hardware and overall techniques to make the wood conversion process available to industry in an equitable manner, that is, on an equal-to-all basis.

Everyone is aware that wood makes lumber, plywood, paper products, and miscellaneous other, more or less common, generally considered unexciting items of day-to-day use. The source of these primary products—the tree—and the timber areas from which it originates are very controversial now. It is nevertheless understood that methods of harvesting our timber resource—our only renewable resource—must change, must improve. Greater utilization of wood fiber must result.

How to approach our goal—total utilization of wood

Visionaries in the timber industry and beyond have long looked at the tree and examined all problems involved in converting that tree into a usable product. They have speculated about the possibility of reducing the tree—totally—to fiber, and then rebuilding it, as Mother Nature did. The fibers would then be specifically aligned so as to reproduce those characteristics needed for lumber and veneer. With those products, it would be possible to build engineered wood structures acceptable to building code requirements. By reducing the total tree to wood fiber, obviously we would approach as near as possible to 100 percent utilization of our timber resource. This total utilization has been a goal of the timber industry for a great many years. This chapter will introduce a process we believe shows that such total utilization and conversion to engineered wood products is on the horizon.

The Fiber Orientation Process

The Fiber Orientation Process developed by WSU does take wood fiber, totally reduced (including mill residues of all kinds), and by specifically aligning the fibers, gives back to the product the kind of structurally engineered properties nature provides.

Products produced on a commercial prototype unit in our laboratory in Portland, Oregon (and products produced on similar equipment at Washington State University), indicate engineered characteristics equal to, and in some cases greater than, the same characteristics understood by us as they relate to lumber and plywood. All types of residual materials, including some of the most noxious elements, have been processed and have achieved what could be considered structural characteristics. Nonmerchantable timber species also have been reduced to fiber and reconstituted by orientation into structural engineered wood products.

Polarization of fibers the key

The process is electrical. An amusing occurrence takes place in our laboratory whenever a visiting group is present. There is inevitably one gentleman who will say, "You can't do what you're doing. I bought a wooden-handled screwdriver 50 years ago so I wouldn't get a shock whenever I was tinkering with the kitchen appliances." What we have been doing does appear to many to be unbelievable or even impossible. We take fiber as produced in commercial refiners. The process we are using is a dry process–not a wet process as is used in making wet-process hardboard or in making paper. We do control and adjust moisture content, for moisture is the conductor which allows us to achieve orientation.

As fibers fall through an electrical field, they are polarized, just as all matter is polarized (Figure 7.1). Negative and positive plates form an electrical field. We polarize the ends of the fiber, negative and positive. The principle of physics–that like poles repel and unlike poles attract–applies. So, unbelievable as it seems, the fibers line up like little soldiers between these plates and are deposited on the mat in what we call full alignment. We use an index which relates to the percentage of the material deposited that is actually oriented through the depth of the mat. This term

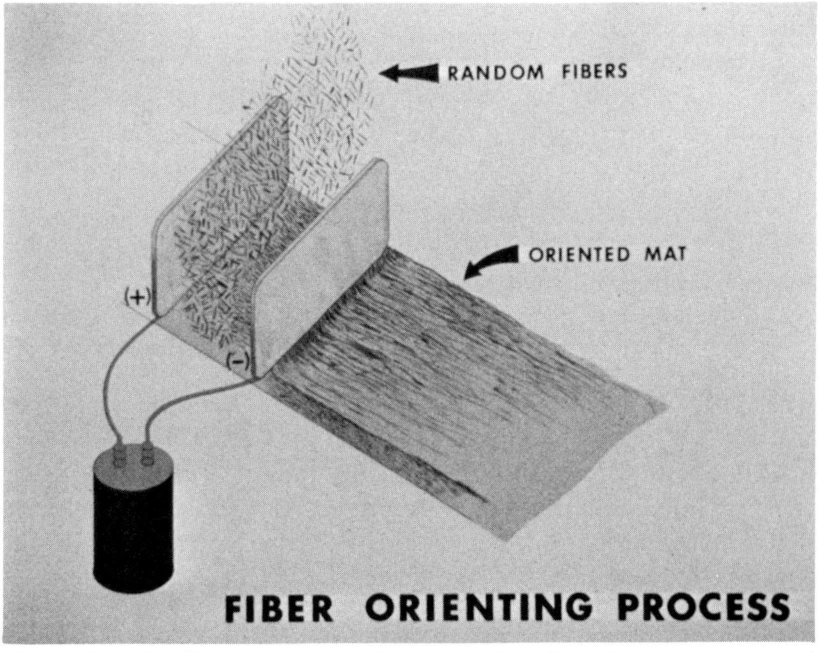

Figure 7.1. As fibers pass through an electrical field between two metal plates, the ends of each fiber are polarized. Fibers then line up between the plates and are deposited on mat in full alignment.

and many others that we have developed are creating for this new activity a language of its own.

Three critical factors are involved

There are three critical factors involved in achieving alignment of the fibers.

1. Slenderness ratio.
2. Moisture content.
3. Electrical power.

The *slenderness ratio* is the length of the fiber versus its thickness or cross-section. We talk broadly of needing something on the order of 10 to 1. Before it reaches the electrical field, the material is, of course, impregnated with the adhesive normally used in particleboard and fiberboard. *Moisture content* of the fiber will vary according to wood specie and whether direct or alternating current is used.

In our laboratory, we use 60-cycle alternating current as our *electrical power.* We are, however, experimenting with both direct current and low frequency. With the capability to use these three variables in electrical power, we achieve many unusual results with our laboratory equipment.

Power consumption is low

The process really is not a great mystery. Applications of electrical fields are used in many ways. They are, however, somewhat misunderstood. For example, a frequent question is whether the application uses a tremendous amount of power. That is a logical question. In truth, there is no current flow in the field, although there is current drain. In projecting a feasibility study for a return on investment for an oriented fiberboard plant over a year, we cannot calculate the amount of electrical consumption involved in the orientation process. (We consulting engineers have a bad habit of using fudge factors, anyway. And the electrical consumption falls well within our normal fudge factors.) Thus the electrical power used is very, very small. Voltage used is on the

Top: Conventional mat forming of particleboard without fiber orientation. Wood fibers have fallen in random directions. *Bottom*: The same forming station, but with fiber orientation. Fibers are aligned lengthwise as they fall through the electrical field and onto the mat, using very small current from the positive and negative plates positioned above.

Closeup of particleboard mat formed with fibers aligned lengthwise. Process produces strength comparable to that of original wood.

order of 100,000 volts. On the surface this sounds like a rather large amount of electrical energy—but in truth, a living room television set uses more than 100,000 volts.

The area affected in a fiberboard plant using the fiber orientation process is the area immediately at the forming station. Stock preparation in a fiberboard plant using the Fiber Orientation Process is just what it would be in any other fiberboard plant; the equipment following the forming station is also the same. The area affected is the forming station where the fiber mat is formed before it is pressed.

Capital needs are low

The cost of making oriented fiberboard really is not any greater than producing unoriented fiberboard. In terms of plant conversion, capital investment depends on the type of former the plant already has and what is needed to fit orientation capability into it.

If the plant is new, the cost will be no greater—and might even be just a slight bit less—if our orientation former is used. So we have not created a monster that suddenly explodes capital costs or operating costs—we really have not affected those areas. There are major improvements, however, in the number, form, and variety of product possibilities with the added strength gained from this orientation of wood fibers.

Market potential is great

Particleboard and fiberboard products, as we know them today, generally are considered nonstructural. The market for these products is, therefore, limited. But with added strength, in a wholly structured form, they can substitute, and probably eventually will substitute, for specific applications of natural lumber and plywood. We go beyond this field, however, into structural applications not involving the general construction industry. Inquiries related to the automobile and aircraft industries doubtless have to do with shortages of petroleum and plastics. They also have to do with the combustible properties of plastics. With structural integrity, formed wood panel products can be substituted for plastic products now used in aircraft and automobile applications. Such basics as dishes and other domestic items can now be given structural integrity by using materials made by this process. Orientation allows a substantial reduction in weight while giving dramatic improvement in structural properties. Many wood products can be made lighter and stronger with all the resultant and obvious savings. We are limited only by our imagination.

Laboratory procedure described

All machinery in the fiberboard plant in our laboratory, with the exception of the orientation former, is just as it would be in any other fiberboard plant. Around the electrical unit, we have a fence to keep people from getting shocked. A continuous nonconductive belt conveys the fiber through the machine. Electrical charges range from 50,000 to 100,000 volts, depending upon the moisture content and the slenderness ratio.

The effect of the electrical field on the fibers is instantaneous.

Electrical panel for laboratory prototype fiber orientation former. Horizontal pipes behind switchbox lines are fiber feed. Fiber infeed box (in background at top) contains brush that separates or "fluffs" fibers before their fall through electrical field onto mat.

Plywood box at left contains vacuum box used to pick up 3-foot-square block of mat for transfer into press at right.

We use a vacuum system underneath the conveyor belt to consolidate and hold the mat down onto the belt. When the vacuum system is on, we get a very fine textured surface in the mat; when the vacuum system is off, we get quite a topography.

Under orientation, festoons of fiber form. The volume of material that we can put through this unit is much greater than can be handled by most formers today. We run the belt to speeds of 90 feet per minute. Above the aluminum plate is a fluffing box, where the fibers are separated to give them individual identity. If the fibers were to remain in clumps or in balls, they would undoubtedly have a neutral geometry and no slenderness ratio. We would, therefore, be unable to orient them. So we break up these clumps so that the action of the electrical field is on the individual fiber. With 60-cycle alternating current, we need between 16 and 20 percent moisture content. With direct current, we go as low as 9 percent moisture content.

With very little effort, we get 40 to 50 percent orientation; with care we get as high as 85 to 90 percent orientation in the mat. A 30-inch by 30-inch board is the largest we can make in our laboratory press.

DISCUSSION PERIOD

VERNON S. WHITE, Western Timber Industry, Portland, Oregon: What is the procedure for someone interested in installing this kind of system? What is the first step?

DORSETT: We would very much enjoy having people come to our laboratory to see this equipment in operation and to see samples of the product. The MK-RDA, Inc. offices are at 14041 N.E. Sandy Blvd., Portland, Oregon. We would appreciate your making an appointment by telephone a few days in advance of your visit.

8

Reliability of computers and scanners in the mill

Dan Pichulo, Marketing Manager, Atmospheric Sciences Inc., Sunnyvale, California

Lowell H. Hulbirt, Manager, Management Science, Publishers Paper Co., Oregon City, Oregon

Part one by Dan Pichulo

This chapter will cover new installations during 1973 of various computer/scanner applications in the sawmill from an operating experience viewpoint. We will examine the reliability and maintenance aspects of scanners and minicomputers in mills to scale logs for accountability, for management information, and for saw control.

At the 1973 Portland Sawmill Clinic, we discussed electronics and minicomputers used together to provide timely management information. But the real emphasis of that Clinic was on how to increase yield by taking better measurements and by using the computer to calculate the best set and then automatically implement it through the saw's set mechanisms.

Many things have happened since. I will summarize briefly.

Automatic quad set

In 1973 we talked about the first quad saw installation to operate with automatic setworks control. It was installed by Publishers Paper Co. at its Molalla, Oregon, mill. A second quad saw was installed by Publishers at its Burney, California, mill. Now, a third quad saw installation is about to go in at a third Publishers mill in Oregon.

Three lumber tally systems were installed during the past year. Log input and lumber output at the same mill can be produced in the same time frame to provide an accurate measure of overall yield and recovery on a daily basis. In the installation phase is a long-log measurement and control system to determine how logs should be bucked. This offers great potential for yield increase. But there is an additional, hidden advantage: the bucking system has the ability to look at every possible cut face and judge whether or not that face has the optimum cutting pattern if the cut were to be made at that point and brought to the quad saw. This arrangement provides the means for two measurement systems to work hand in glove to produce an additional or synergistic yield.

Automatic controls for headrigs

Other events of the past year concern different types of headrigs. We have seen a number of double-taper chipper canters and Chip-N-Saw type headrigs automated for set control. These systems are in the Northwest, in Canada, in the South, and on the Atlantic coast. One mill with fully automatic control with a 150-foot-a-minute Chip-N-Saw has reported an overrun increase of 17 percent.

Another development in the past 12 months–including two major installations–has been automated sorting. In northern Idaho a six-bin sorting system feeding two Chip-N-Saws was installed. Logs entering the sorting area at 400 feet per minute are automatically measured and classified by diameter at that speed and kicked automatically into bins. It's a sight to behold, for it provides productivity increases well beyond manual capability. In Canada, another sorting system operating at about 300 feet per minute sorts logs and actually doubles the production of the mill.

Scaling for management information

Additional scaling systems have been put in for accountability as well as to provide the right kind of management information for better control of the entire mill operation.

I feel the familiarization period for scanning systems and mini-computers has been completed for this industry. We are beyond that phase. We are now moving into the action phase and need to examine how these systems apply to each mill and application. The newer systems now operating are not simply saw control systems—they are multiple-application systems. Versatility has been proven. Operability is proven. But most important, financial justification has been proven. What about the justification on saw control? In most cases it has taken less than 12 months to pay back the investment.

With this progress, there seems to be one more stumbling block still waiting to be overcome before the use of computers and electronics in the sawmill industry really takes off. The potential is here and ready to be used.

Reliability question

But this stumbling block is reliability. There does not seem to be a strong belief in the reliability of these systems overall in this industry. Today I will try to dispel that doubt.

When most people think about computers, scanners, and electronics, they think about little black boxes. And when they think about these black boxes, they visualize them as containing delicate, tiny, fragile pieces of equipment. But these little black boxes are more rugged than you think.

Computers quiet

Computers do a good job and they do it silently. They do a better job than a human being at setting saws and running decks. And they do the job more consistently, do it more reliably, and they do it without any verbal feedback! With a little planning and foresight, nothing is going to stop that computer from doing what it is supposed to do.

If that is the case, why aren't there more computers in use in mills? I would have thought that by this time the industry would have been inundated with electronics. So I have tried to categorize the questions I hear people asking about the use of electronics and

After barking, scanners sense each log's diameter on the quad saw infeed conveyor. Information is instantaneously fed through a computer, which instructs setworks where optimum cuts should be made.

minicomputers and scanners in mills. I would like to share some of those questions with you now.

The most frequently asked question from people on the verge of deciding to install a computer/scanner system is this: "Why do you think the penetration of electronics and minicomputers and scanners into sawmills hasn't been faster?"

Wait-and-see attitude

My response to that question must be that the rate has been reasonably fast, provided one considers a small group of companies like Weyerhaeuser, Publishers Paper, and other companies in that group. In fact, the penetration rate has been extremely fast for these companies. The rest of the industry, however, is still waiting to make sure that the sixth or seventh unit put in by these companies is going to work as well as the first one—or that the first installation still will be working when the seventh one goes in.

I am inclined to believe that the penetration rate has been slow among other segments of the industry based more on the psychology of the situation than on the facts of the situation. People have been preconditioned to believe that computers are bad news.

It is difficult to change attitudes. A German philosopher once said, "If you tell the public something often enough and long enough, it's going to become fact." It's going to get into the subconscious mind and every time the rational process of decision making tries to handle it, it's going to be there gnawing away saying "no, no, no, no."

Mental holdover

I would like to conduct an experiment to emphasize that point. Please complete the last words of these lines:

Pichulo:	Pepsi Cola hits the . . .
Audience:	. . . spot
Pichulo:	Twelve full ounces, that's . . .
Audience:	. . . a lot
Pichulo:	Twice as much for a . . .
Audience:	. . . nickle too
Pichulo:	Pepsi Cola is the drink . . .
Audience:	. . . for you

Good, you remembered. You heard it on television this morning, that's how you remembered it–right? Wrong! That jingle has never appeared on television and it has been 23 years since it has been on radio. Now I am sure you have not been thinking about that jingle for 23 years! But it is the same principle with electronics and computers. "Computers are trouble; computers are difficult." These electronic tools work for some companies–and these companies do not have a monopoly on brains. These companies make them work, and so can you.

Reactions varied

The second question I hear at this point is, "Why is there such mixed and sometimes extreme opinion about minicomputers and scanners, even within one company? One person likes them; another does not." Many of you may identify with that question.

I think the problem is that computers require people to think logically and sequentially and utilize their ability to analyze interrelationships. How will change in one area of the plant affect another area? Only then can men put their intelligence inside the computer so that the computer can use that intelligence.

Of course, there are people who think, people who think they think, and there are people who would rather do anything *but* think. One way to counteract these negative opinions is to listen to reports about what other people are doing with computers and about the yield increases, productivity gains, and already achieved returns on investment.

Soon we will overcome that little stumbling block. It will disappear and computers will be everywhere doing the kinds of jobs they can do best–the repetitive jobs, the routine jobs, the

Banks of computer electrical connections are tested by employee of Atmospheric Sciences at its Sunnyvale, California, factory. The firm is a pioneer in application of scanner/computer techniques to sawmills.

boring jobs–because they can do these jobs better than any human can.

Here is a particularly important question that occasionally is brought up: "I've heard all this talk lately about management control information. That's really a bunch of nonsense–it's nothing but a fancy name for a typed piece of paper. There's no information anywhere that is as good as what a manager sees happening in his own plant with his own eyes." I imagine many of you identify with that question. And there is a valid point here.

Probabilities mistaken

However, computers were never intended to replace the manager's ability to go into the plant and understand the operation.

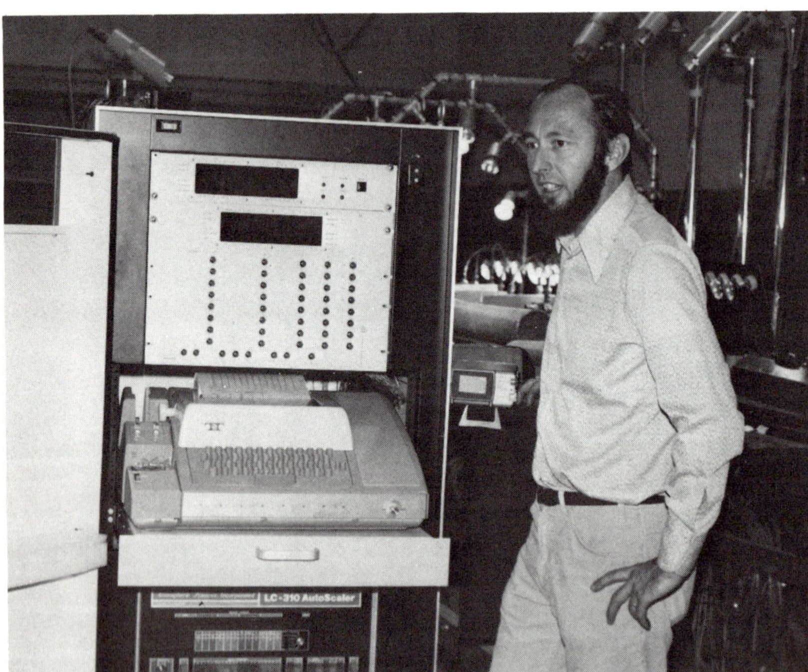

Readout of the computer's decision is available to sawyer, who can override it. Several programs (dimension, shop, full-sawn, etc.) can be stored by the computer. Information on a shift's performance can be accumulated and read anytime during the shift on the system's automatic teleprinter.

But being in the right place at the right time to observe the wrong thing happening is a matter of probability. The odds are against it. How can a manager take a 10- or 15-minute trip through his plant, look at all the operations, all the pieces of moving equipment, all the pieces of wood in various states of disarray and in various stages of breakdown, and then integrate that information as to yield, overall productivity, and total effectiveness of the mill on that shift? The human mind has great difficulty integrating so many factors simultaneously. It is lucky to handle two or three factors at the same time.

Let's prove that. Will everybody in the first eight rows please stand up? Now, we have about 50 people standing and we can do a brief experiment. What are the odds that any two of those standing have the same birth date—were born on the same day in the same month? What are the odds for a group of 50? Would you say 5 to 1? Or 6 to 1? Who will bet me even money? I'll bet that two of them do have the same birth date.

If you were born in January, please raise your hand. We have three in January. What date, sir? The 29th, the 15th, and the 25th. OK, please be seated. Will those born in February raise their hands? Look at this. The 12th, the 17th, the 6th, the 6th—got a match already—the 1st, the 12th, the 12th—two matches—the 6th—three matches. Please be seated and thank you, gentlemen.

Someone says it's rigged? Sure it's rigged, but not in the way you think. The odds are in my favor. The odds are about 23 to 1 for a match with a group of 50 people. It may be difficult for some people to believe this, but it can be shown mathematically to be true.

It is the inability of the human mind to integrate multiple operations that prevents it from really measuring the true effectiveness of an operation. This is an area where management control information can pay big dividends.

Reliability and maintenance costs

Next question: "Is there any way I can be assured of a high level of reliability? And what about maintenance cost?" I think everybody can identify with those questions.

At least one measurement and control company has a staff of

factory-trained people located in strategic geographical areas to provide service on these systems as needed. I understand that this company also offers annual service contracts to guarantee fixed costs on both preventative maintenance and as-required services and repairs. It is reassuring to know such service is available.

Mill experience the final proof

Another question: "All this business about reliability or unreliability is very hard to come to grips with intelligently. How can I put this issue in its proper perspective?" I think one of the best ways to answer that question is to hear from someone who has lived through it—and particularly, someone who lived through it from the early days. It is important to hear from somebody who started at ground zero and worked up to today.

Lowell Hulbirt is with the Times Mirror Co., the parent corporation of Publishers Paper. Lowell has an unusual title—he is manager of management science. He not only looks at the big picture for Publishers Paper, but actually creates it and makes it happen. The key word one might use to describe his experience is not *theory*, but *application*—application of control systems to processing operations. Lowell will share with us some of his first-hand experiences with computers and scanners in the mill.

Part two by Lowell H. Hulbirt

First, I will cover a few general aspects of maintenance with computer/scanner systems. Then I will discuss specific experiences of our company—Publishers Paper—in the area of computers and scanners in the mill.

Discussion will center around the original control system at Publishers Paper Co.'s Molalla, Oregon, sawmill. We have a similar installation at our Burney, California, mill. Both installations are for control of quad band headrigs. Maintenance experience has been slightly different for each. We have had many more problems with the initial installation—I think because it *was* the first one—than we have had with the second. So that, in its own way, is somewhat encouraging.

Mill environment problem

The first thing we discovered was that the main problem (or the cause for whatever little problems there were) was mill environment. The control systems, of course, are designed on the drawing board and in the lab. No one is really sure exactly how the system will act when it is put into the mill. In terms of maintenance problems, the factors we found to be important are those which exist in almost any sawmill.

First is the matter of dust in the air. I am talking about three or four kinds of dust. We have a lot of sawdust floating around; we also have bark dust and yard dust. The dust problem, of course, is more extreme during the dry season than it is in the winter. There are also problems with moisture. A third source of difficulty is vibration, which occurs when the sensor or the control system is located close to another major piece of equipment, such as a debarker or a saw. Much vibration is encountered in any mill, in some more than in others. It ranges from that caused by the operation of heavy equipment and large motors to the constant pounding that the whole system—conveyors and everything else—gets from having several thousand logs per day dumped from one level to another—onto the log decks and down to the conveyors.

Outside light affects reflective-type scanners

We also found that we had problems with outside light. We are using mainly reflective-type sensors, which depend upon reflective light, although we have some of other types, also. We found that we had to be very careful about where the reflective-type scanners were located, how they were focused, and how they were shielded, because of the outside light problem. Here again, we found we had a seasonal problem. In the winter, when there was not as much sunlight, we did not realize what the problem was; but in the summertime, we experienced periodic misreadings of log dimensions due to the effect of the outside light.

Then there is the general problem in the sensing area of controlling debris—that is, chips, bark, or any kind of debris which falls in with the log and has to be taken away, blown away, or removed in some manner.

Here are a few things which we experienced that other sawmillers will probably be able to relate to their own mills. These are examples of the kind of maintenance which is involved with an electronic control system and they point out some things to be on the alert for. This will not be a technical discussion. These simply are the kinds of things that can happen that will shut down your control system until you correct them.

Log tally inaccuracy corrected

In one case we had a peculiar condition where we were getting good log measurements but all of a sudden started getting a log tally from our saw which was two to three times the actual number of logs being put through the saw. We traced through the electronics and found they were working correctly. After considerable searching, we finally discovered the cause of the problem. A holddown wheel on the top of the log had gradually lowered itself until, in idle position between logs, the tire on the wheel was breaking the light beam which tallied the logs. Every time the holddown wheel finished with a log, the wheel would bounce down into the light beam and be counted two or three times. When we found out what the problem was, it was easy to fix. It was a routine maintenance task to adjust the idle position of the wheel and correct the misreading.

I-beam shadow discovered

In another case, we installed a new log deck adjacent to the quad saw and a couple of things happened. After installation we found out that log measurements were not changing each time a log went through the saw. We seemed to be stuck on one dimension for long periods. Finally the computer would register a new measurement but then stick on that measurement. Well, it took a lot of searching, but we finally found that a shadow from one of the I-beams which had been installed in connection with the new deck was falling across the program sensor which cleared each log out of the computer's memory and said, "OK, get ready for the next one." Once we discovered the cause, it took a maintenance man with a torch only a few minutes to cut off a little edge of the flange that was casting the shadow and we were in business. That is an example of the kind of maintenance work that can be needed when using electronic sensing devices.

Sensor burnout solved

In another case, one of our sensors burned out very suddenly and unexpectedly. The sensors had been doing a good job and there was no particular reason why this should have occurred. So we checked the electronics again and could not find anything wrong. The voltages seemed to be all right. Everything was under control. Then we started looking around the area and noticed that a particular sensor was in line with, and close to, a new catwalk, just installed. We did not see it happen, but we are pretty sure that whoever was welding to put up the catwalk was just not aware that a sensor was there, so it was exposed to welding flash, which burned it out. A simple, ordinary thing, caused by lack of awareness on the part of mill people who really had nothing to do with the control system itself.

Focal point shifted

Another very instructive case was brought about by an improvement we made. We installed a new log rail on the infeed side of our quad saw. It was a needed replacement—the old one was well worn down. Then we noticed that, suddenly, the logs going through the saw seemed to be getting larger. We were showing diameters about an inch bigger than we had been processing previously. Well, we really could not understand why that should be, since we were running the same mix from the yard as before. On checking, we discovered that the new rail on which the logs were riding was about one inch higher than the old one. Our diameter scanners had not been readjusted and were still in the same position—focused for the old log rail. Their focal point was one inch below where the log was, and as a result we were measuring every log too large. Again, a case of something fairly simple; a problem caused by lack of awareness on everyone's part that there was a measuring system involved and that it was related to other equipment in the mill.

Another problem I would like to mention is one which is common, I think, to any sensing system. We do get inaccurate dimensions as we go along—not every log is read perfectly. One reason is chips on the rail. In most areas where the log comes into the saw, there will be chips, and it is easy for chips to pile up on the framework and infeed devices (we have a chain). It is a lot of bother for someone to go out there and keep cleaning chips from the area under the sensors. So we find that periodically we are reading a chip sticking out beyond the end of the log instead of reading the log itself. If you do not keep the area clean, optical sensors will not be able to distinguish between the log itself and other wood in the sensing area.

Infeed chain measured

Another thing that happens to us once in a while—we are not too surprised by it anymore—is that periodically we get a small-end diameter reading of 2.4 inches. Initially we were really puzzled because we do not run logs that small. The large-end diameter was correct, but the small end was recorded as 2.4 inches. It took a while before we discovered that 2.4 inches was the width of the infeed chain, not of the log. Some of the light from the spotlights in the system had been bouncing off bright spots on the chain and reflecting back into the sensor, so the sensor was reading the width of the chain instead of the log diameter. Correcting this problem was simply a matter of adjusting the lights and watching for bright spots which could cause errant reflections into the sensing system.

A very common thing—and this does require daily attention—is the effect of dust and moisture. You must have regular cleaning procedures. You have to wipe off the lenses. You must keep the light bulbs clean. You have to maintain the whole system so *it* can see as clearly as the operator can.

Good housekeeping a must

This brings us to what I would classify as the single most important point about the maintenance of a computer/scanner system. It is not an electronic consideration at all. It is a matter of good housekeeping. The system is sensitive enough to require an effective cleaning system. It may involve blowers or someone physically cleaning, but whatever the method, the operation must be kept clean enough to give a good reading of the log.

Dust has been the worst single problem we have had at Molalla. This is because of the particular locations of the computer and of the electronics. Due to the existing mill layout, they had to be located very close to other dust-producing operations. At our Burney mill we had a different situation. There, we had a large cab with a separate computer room in back, providing a double airlock in and out to protect the computer from the mill atmosphere.

Dust excluded

At Molalla, the computer is in a cubicle of its own, which is filtered and air-conditioned. But the door to that cubicle opens directly toward a debarker, six or seven feet away. Every time somebody goes in to take a report or check readings on the computer, a lot of moisture, bark dust, and sawdust is sucked into the computer compartment. Eventually this builds up inside the computer itself. The computer is a pretty rugged piece of equipment. You do not have to be afraid of having one in the mill. But dust is one of the things it does not like. We were not even aware of the nature of this problem, or the extent of it, for the first six months. The computer worked perfectly and was the most reliable part of the system. Our problems were with the other equipment, the optical and scanning portions of the system.

But eventually the computer started giving implausible readings and we called in an outside service man, as we are not computer technicians. He checked inside the computer and found there was enough dust inside to produce shorts and overheating between the panels inside the computer and in the memory unit. He vacuumed, plugged everything back in, and we were off and running again. That was all it needed.

Well, the first time was not too bad, but the same thing happened only three or four months later. We were getting a build-up; the technician could not get it all out. Also, we had entered the dry season when dust conditions were worse than before. So this problem became more and more frequent. We learned that we had to have periodic, routine cleaning. That includes not only cleaning the room, but vacuuming the inside of the computer about once every 30 days. Now that we have learned this, we have the problem under control; but it took some experience and some downtime. This experience led to development of a separate cabinet—a so-called environmental cabinet—which we now use in new installations. Here, the computer and the teletype printer are housed in a separate cabinet which can be within the operator's booth or cubicle. This allows two filtering systems—one, the booth itself, and two, the cabinet for the computer. We hope this double filtering will eliminate the need to vacuum every 30 days. But we will be alert to the possibility and will continue routine maintenance attention.

Two options available in approach to maintenance

A word about our general approach to maintenance. In setting up a maintenance program, there are two basic options: (1) obtain a service contract and have someone come in and service your installation whenever you need it; (2) train your own mill personnel to perform the necessary maintenance.

We chose to take the second route. We wanted to train our own mill people. This was not necessarily to save money—a service contract costs, roughly, about 10 percent of the installation cost per year. Our main reason for this choice was to have help available when we needed it. If some little thing went wrong, we did not want to have to call, then wait for a field man to get there a day or two later to fix it. We wanted our own people to be capable of doing some troubleshooting.

We do not attempt to do any of the technical electronic work. We are not computer experts. We do not work internally on the computer or the sensors or scanners. We call in outside help from the supplier for service of that sort. But we try to do the other type of maintenance—the kind I have been talking about—ourselves, along with general troubleshooting on the system to spot what is causing a problem. To be able to do this, you need to select someone on your maintenance staff (probably a mill electrician who has an aptitude and an interest) and give him some special training. The most technical thing I have mentioned is cleaning inside the computer. To do this, it is necessary to remove some electronic gear and put it back. If this is not done correctly, you will not have any program, or any control. The program must be read back into the computer and the computer restarted after cleaning. Training is required to do this job properly.

Maintenance responsibilities divided

Our general division of maintenance tasks is as follows: we maintain the physical setup, but call in an expert from the equipment supplier as needed to service the computer and the internal electronics.

The question of reliability was raised previously. There are many opinions as to the reliability of these systems. We have had our share of troubles; but we have also had our periods of very good operation. I do not know exactly how one measures reliability. It can be measured from many different standpoints. To give a general idea, it might be helpful to talk about the amount of uptime, versus downtime, we have had. Since we started the first installation—a period of two years' experience—I would generally estimate that at Molalla (our initial system) uptime has been on the order of 80 to 85 percent. The percentage has been better than that at Burney, our second installation—uptime there has been about 90 to 95 percent.

Do not be misled by the fact that the system has few moving parts. This does not imply that the system will be 99 to 100 percent operative. It does not happen that way. The system requires maintenance to keep it going—particularly in terms of external sources of trouble. In general, you should plan about the same amount of maintenance on this kind of a control system as you would plan for any other piece of equipment costing in the \$50,000 to \$75,000 range. It is not mechanical-type maintenance; it is housecleaning, troubleshooting, and electrical maintenance.

Training the crew

A good maintenance program should certainly include two basic things: first, as much training as you can give your mill people in the workings of the system (and that will take some time away from other things). Training sessions preferably should be with someone from the supplier or someone who has experience with the system. Second, do not forget to keep a supply of spare parts on hand. These parts often cannot be had on a moment's notice. Some are special parts. I compare it to buying an airplane. If you do not buy the spare parts with the plane, you are not going to have them on hand when you need them. So these two elements—training and a supply of parts—are essential.

DISCUSSION PERIOD

QUESTION: If you have to call in a factory man and the operation is down for a day or a day and a half, how long will it take to recover the lost production?

HULBIRT: We do not have mill downtime as a result of computer failure, because we have a flexible system. We have both a manual system and an electronic system. So whenever the electronic system goes out, the manual system takes over. The operator is not getting the same amount of information as to the size of the log and the setting of the saw as he did with the electronic system, but he can still push the button and operate the saw. So we avoid any downtime caused by computer failure.

As to the length of time it takes a service man to get to the mill when you call, fast service would be the following day. The technician generally is not located in the immediate vicinity of your mill. He may be on some other job. Generally he will appear one to two days later to fix your problem.

LONNIE BAIN, maintenance superintendent, Publishers Paper Co., Molalla, Oregon: I would guess that the maximum time we have had to wait was three days, when the technician was on an out-of-state job. One day is normal. Often it is the same afternoon. In fact, the technician will come in at 2 o'clock in the morning to work on the problem. They do that often, as they like the mill to be down while they are working on it.

QUESTION: Mr. Hulbirt, I hope I got the wrong impression from what you said about downtime. I believe you said that uptime at Molalla was approximately 85 percent. If that is true, does that mean the computer is out one day a week?

HULBIRT: Well, that does not quite add up to one day a week, but it means we have periods when the computer is not working. Usually, our experience is that when it is working, it is working very well. Then something may go wrong. It may take a couple of days to find out what it is—like the simple little things I mentioned. The computer is not always the cause of the problem. In many cases it is something in the outside system. In terms of uptime, I am referring to the system as a whole, not the computer itself. The computer has a much better record than that. Downtime on the computer has been pretty well limited to those failures that were corrected by cleaning.

QUESTION: Has mill downtime decreased since installation of the computer?

HULBIRT: At Molalla, downtime decreased substantially after the first couple of months. There was a period, like a debugging period, during which the computer settled down into its operation. Then it operated very well for about a year and a half. Now we are beginning to have additional problems. I understand from talking to the factory representative that this is not unusual. A computer has a life cycle. It does deteriorate; it is not good forever. We are talking about minicomputers under mill conditions. This is a much tougher installation than the centralized type common in most accounting departments or an office environment. So in the mill you have a good running life of somewhere around two years. Recently we exchanged our original computer at Molalla for a newer model. Part of the reason was because of deterioration. We were having trouble with the power supply unit in the computer. Rather than repair the unit, we went to a later model, which has some features the first one did not have. We exchanged—traded in—the old one. It did not cost us anywhere near the price of a new one.

QUESTION: What is your estimate of cost savings, compared with the previous manual operation?

HULBIRT: We do not give out exact figures, but I can tell you this with great assurance: we more than saved the cost of the entire installation during the first year of operation. That is a conservative statement.

QUESTION: When you compare manual measurements with the

electronic measurements, what kind of accuracy do you experience? In other words, what is the difference in accuracy between manual and electronic measurements?

HULBIRT: There is a great deal of difference. When we talk about manual measurement, we are talking about the operator's judgment, a rather instantaneous judgment, as the log comes into his view. Is that what you mean?

QUESTION: No, I mean: to check the accuracy on your system, you probably make manual measurements. How do these compare with the electronic measurements?

HULBIRT: Yes, we do make manual measurements, and we check both the diameter reading and the length reading. The length readings are not nearly as significant or as important—we are worried only about 2-foot increments in length, although the system measures length within a couple of inches.

Diameter is extremely critical, and we do check this closely. This has been one of the most reliable portions of the system. Diameter readings are accurate within 0.2 inch, on the average. That degree of accuracy has been maintained from the original installation.

QUESTION: How often do you check your computer readings against the manual readings?

HULBIRT: This is part of routine maintenance. What would you say, Lonnie?

BAIN: We check lengths and diameters weekly. We also check the log count. About a year ago, when we first got the system going, management was complaining that the tally on the green chain or on the sorting system was a lot less than what was coming out of the computer and what the operators had been marking down for the quantity of logs that went through the quad saw.

So I went up into the computer room. It has a little window that looks out into the operator's compartment. The operator has a manual counter with which he tallies the logs in. I put the computer on scan, so it would print out a description of every log that went through. Then I sat there and watched the operator. A foreman would walk by or a log would get crossed up and the operator would forget to hit his little lever, so the computer actually produced more reliable information than the operator did when tallying the logs manually.

QUESTION: Assuming you maintain perfectly clean conditions for the computer, how often would you estimate it would be down because of mechanical failure of the computer itself? How many times a year?

HULBIRT: Well, that is tough to answer. The only downtime we had on the computer during the first year and a half was caused by dust. Aside from that, it did not give us any trouble—until recently when the power unit did start to fail and we replaced it.

QUESTION: Would you regard that as an average?

HULBIRT: I would not want to suggest that. I am just saying that that was our experience. I am sure others would have a different experience. We will probably find our situation at Burney much better than at Molalla. Molalla has tough mill conditions—about as difficult as we will get. We know now what to provide against. Also, the later model computers are designed with mill conditions in mind. We expect the new model to perform better than the first one did.

QUESTION: Is it necessary to install the computer in close proximity to the cutting station, or could you have the control some distance away?

HULBIRT: The computer does not have to be in close proximity, but there are some advantages. If you locate your computer in an office building, it will cost you something to run the wiring to a remote point and back again. In addition, it is very convenient to have the computer right there when you are doing checkouts. As Lonnie Bain mentioned, we have a computer cubicle where we can print out the dimensions of each log. You can check that against the log as you look at it and as it goes into the saw. You can also check the saw setting, which shows on the operator's console, at the same time. That is why we have a window between the computer and the operator.

DICK SETTERSTROM, State of Montana, Butte, Montana: What is the applicability of this equipment to smaller operations?

PICHULO: My answer is, of course, biased; but I think there is no question that any mill can benefit from use of such a system. The question of applicability does not depend on the size of the mill. What we are talking about is upgrading mill practice.

It is clear that in a small mill where you do not have as much volume you are not going to have the same amount of payoff from upgrading volume. But it may be just as important to the profit picture. What you really have to look at is the amount of investment that must be made. It is in the $50,000 to $75,000 range for an individual control system for a given saw. If you are willing to spend that, you will get it back. You would have to apply the improvement factors to your own volume to determine the length of time required to get back the equipment investment.

9

Setting a precision carriage by computer

John D. Steadman, District Sales Manager, Filer & Stowell Co., Beaverton, Oregon

We all are interested in finding ways to improve production techniques, to make them more efficient, to produce maximum yield and profit from our forest resources. The key words to remember are *resources, efficiency, yield,* and *profit.* Everyone profits from more efficient, higher yield utilization of any natural resource.

Many products and systems have been developed in recent years to help achieve this goal. In this, and previous Sawmill Clinic volumes, we have learned of the practical application of: (1) log scanning devices; (2) log merchandising systems; (3) high-production small-log sawmills; (4) thin-saw edgers; (5) high-yield log breakdown patterns; and many more products and systems for attaining maximum yield from the raw material.

As more and more new systems go into operation one critical fact keeps coming back into focus:

> As long as we have logs of any size available, as long as grade is a factor, and as long as trees remain nonuniform, there will be a need for a headrig breakdown system consisting of a bandmill, carriage, log turner, feed mechanism, and a skilled sawyer to make it all work.

This chapter will discuss recent developments in carriages and carriage setworks and their effects on lumber yield from logs.

The history of carriages

Let us look at some history of our industry. Carriages themselves have evolved through stages of hand dogs, to air-operated hand dogs, to remote sawyer control. The knees have been driven by rack-and-pinion, screw-block, chain-drive, and hydraulic-type movement mechanisms.

Setworks have been powered by alternating current electric motors, air motors, and hydraulic motors, typically through a gear reducer and a chain-and-sprocket drive to the carriage set shaft. These were first operated by setters on the carriage, then by remote sawyer controls.

Position control, or setting, has been by limit switch, potentiometer balance circuit, or by counter-type control system. In each case it has been an *open-loop control.* That is, the sensor, limit switch, potentiometer, or counter pickup senses when a given position is reached. It passes a signal through a relay to cut off power to the motor and to set the brake. After that, it is up to the brake to stop and hold the knees. The sensor has no further control of the operation.

Accuracy potential of conventional equipment

A typical conventional carriage and setworks will function to an accuracy of about plus or minus 1/32 inch *most* of the time—and if all the mechanics are properly maintained. This is the most that can be reasonably expected. Age and/or lack of maintenance will cause deterioration from this level of accuracy.

Brakes, clutches, chain drives, rack-bars, nut blocks, and so forth, are all subject to heavy wear. Potentiometers, limit switches, and brakes have also been known to drift—particularly during the first hour of operation after a cold night. All these elements contribute to the plus or minus 1/32-inch accuracy under certain conditions due to wear and tolerance buildups.

On top of these mechanical and electrical hardware factors, is sawyer judgment. Typically, a sawyer reads his backstand dimension from a mechanical dial, reading to within about 1/4 inch, maybe 1/8 inch, with a sharp eye. From this reading, he must mentally calculate his runs, allowing enough surplus on each cut to prevent cutting any thin boards but not so much as to produce shims. A skilled sawyer is quite a calculating machine when you get right down to it.

Products developed in the last few years have virtually eliminated the wear parts that create slack, the variable time-acting parts that create wider swings in accuracy, and the mental calculations and guess-work involved in evaluating the backstand available and in setting up a run that will make best use of it. What are these new products? How do they compare to the older systems? What makes them better? What results can they produce? And, perhaps the biggest question of all—are they worth the price?

What are precision carriages and setworks?

The definition of a precision carriage might go something like this: "One that eliminates as many as possible mechanical linkages and drives, that uses antifriction devices wherever possible, and that—where friction cannot be avoided—uses long-wearing renewable or replaceable low-friction materials."

The definition of a precision setworks might be as follows: "One that gives an accurate indication of backstand distance at all times, that allows accurate set adjustment (not cut-and-try), that assures positive repeatability of sets *all* the time, that gives the flexibility of an easy change-of-set function, that eliminates the need for mental calculation of runs, and that reduces to near zero the chance of heavy or scant boards and shims."

The precision carriage—a detailed description

A typical precision carriage drive train (Figure 9.1) consists of the set shaft, a single-reduction precision bevel gear, and a ball screw-and-nut combination. The nut attaches to the knee much like the older style rack bar in a rack-and-pinion-type carriage. These elements provide maximum mechanical efficiency, minimum friction, and, as a result, long life with little degrading of performance. The sensor, or pickup unit, is direct-coupled to the

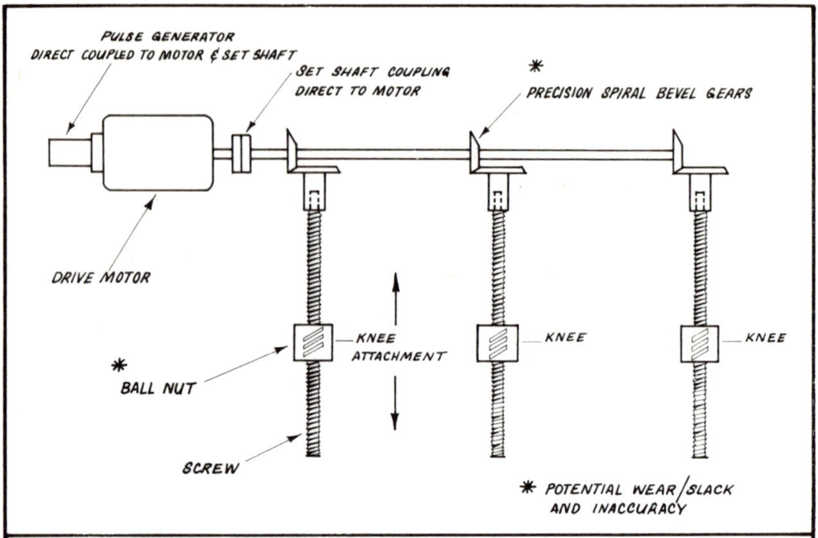

Figure 9.1. Design of drive train for a precision sawmill carriage eliminates as many potential wear points as possible.

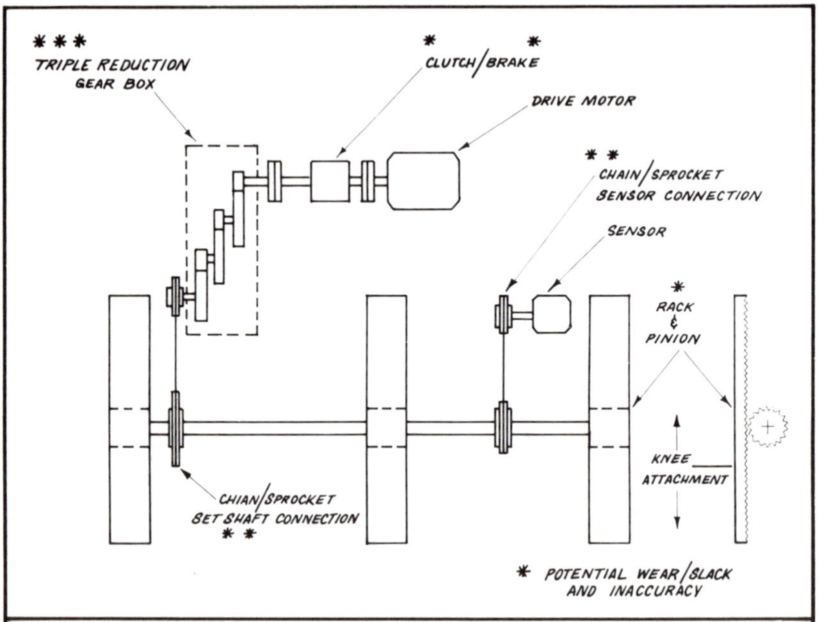

Figure 9.2. Usual elements in a conventional sawmill carriage drive train. Sensor is two mechanical links from the knee face.

set shaft—in this case, through the motor.

Compare the precision drive—in terms of simplicity, efficiency, and maintenance—to a conventional drive train (Figure 9.2) with a clutch/brake, multiple-reduction gear reducer, chain-and-sprocket drive, set shaft, and rack-and-pinion or screw-block mechanism to move the knees. The sensor is coupled to the set shaft through a chain-and-sprocket drive. So we have further complicated the mechanics—the sensor is two mechanical links from the point we want to control: the knee face. That gives a rough idea of the basic mechanical drive differences between precision and conventional carriages. Now, let us look at some refinements.

The knee-to-headblock mounting (Figure 9.3) is of course important. Low-friction, long-life wear strips under the knees assure continued square knee-to-headblock relationship. Headblock wear strips keep the log abrasion away from the knee bearing surface. Replaceability of these items makes reconditioning easy, as wear does occur. To get full value from all these features, we must be

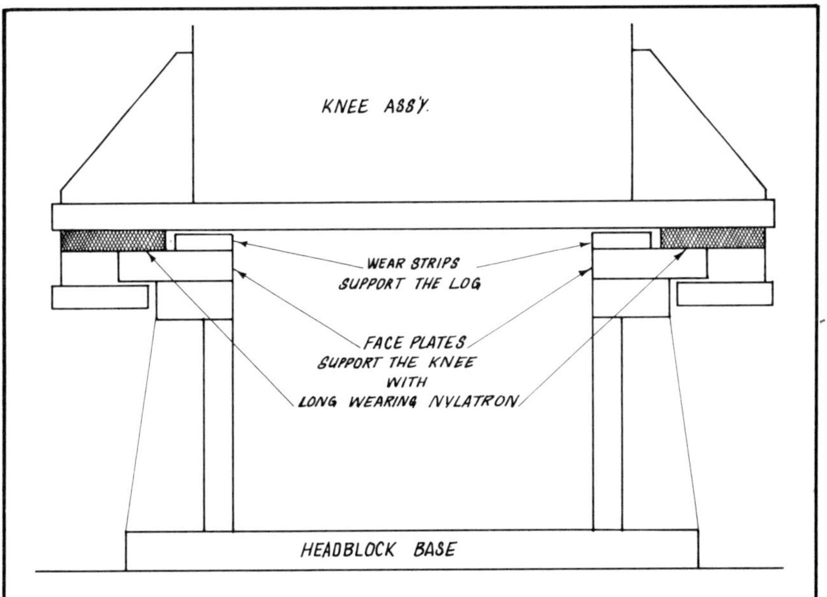

Figure 9.3. Knee-to-headblock mounting on a precision carriage features wear strips under the knees to assure longer accuracy.

sure that all the headblocks and knees are in line and working together.

A solid structural steel frame (Figure 9.4) is the foundation of the complete carriage assembly. Machined mounting pads, top and bottom, for headblocks and trucks and corresponding machined headblock and truck flanges assure proper alignment. This combination of machine features mounted on well-machined and leveled rails on a solid foundation will provide headblock and knee alignment to within better than 0.005 inch.

The precision setworks

The ideal setworks to operate with this machine uses a closed-loop servomechanism type of control—a fancy term, but simply explained by a comparison with a conventional, or open-loop, system.

In the open-loop system—such as the limit switch, clutch-brake combination shown in Figure 9.5—the limit switch opens when its preset position is reached. It then trips a relay, or solenoid valve,

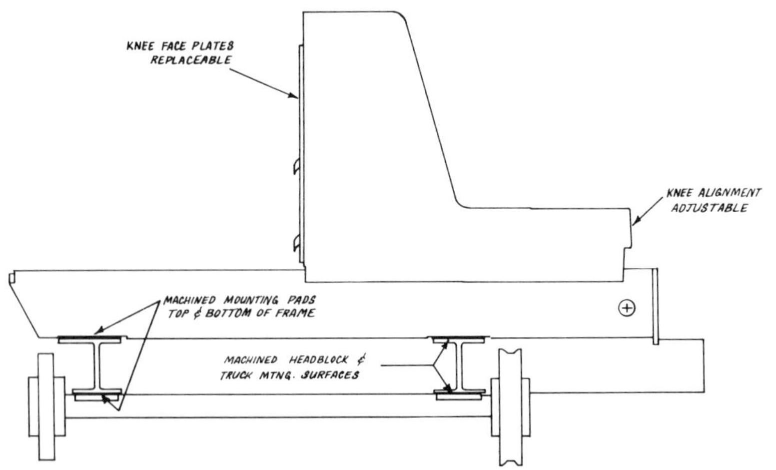

Figure 9.4. Solid-steel frame truck mounting provides a foundation for a precision carriage assembly.

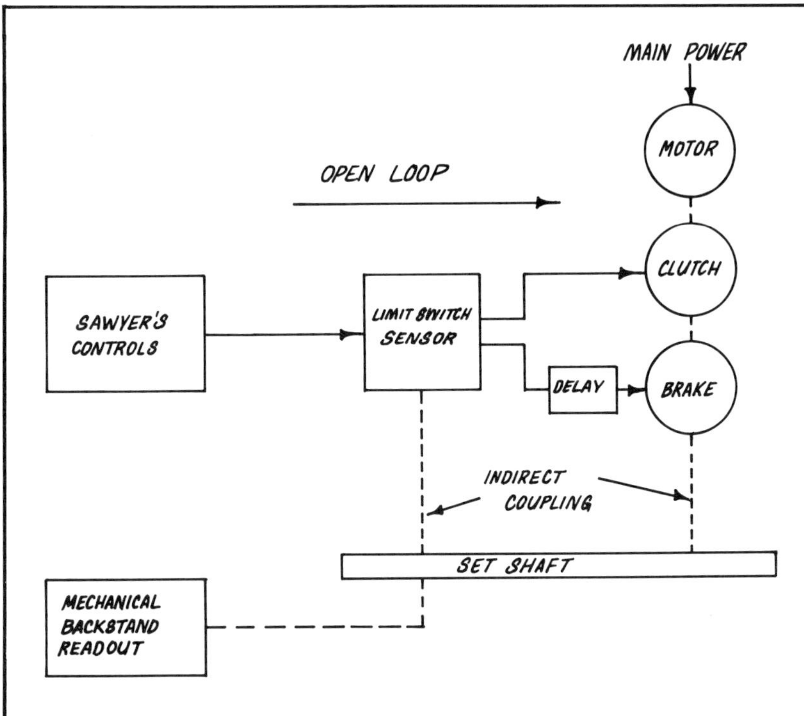

Figure 9.5. On the conventional, open-loop setworks system shown, areas of "coasting" and play after the switch is thrown make it impossible for the sawyer to know exactly where the knees do stop.

or multiple relays. These, in turn, disengage the clutch and set the brake. The relay, solenoid, and clutch-and-brake action all take place after the switch is tripped. Neither the switch nor the operator knows for sure exactly where everything came to a halt. The mechanical backstand dial gives only a rough estimate of backstand dimension.

Closed-loop system

In the closed-loop system (Figure 9.6), the position sensor *knows where it has to stop.* It operates in anticipation of that stop point, controls deceleration of the system, and stops at the point that was selected. It then *holds* position against any external force

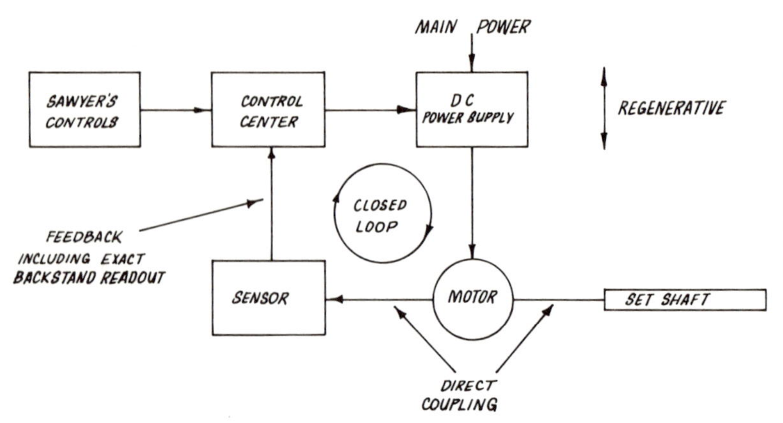

Figure 9.6. In a closed-loop setworks system, the position sensor stops on target and holds its position, assuring accuracy.

that tries to move it. About 0.001 inch of knee movement will produce maximum countertorque from the drive system to counteract it. The sensor, or pick-up, is a pulse generator directly coupled to the motor shaft. It gives the sawyer an exact digital readout of backstand dimension at all times.

Controlled deceleration is possible through use of a direct current drive motor with a regenerative solid-state or SCR (silicon controlled rectifier) power supply. *Regenerative solid-state power supply* simply means that the direct current motor, when decelerating, becomes a direct current generator and pumps its energy back into the supply line. In doing so, it brakes itself to a stop. The motor then serves as its own brake. No mechanical friction-type brake is required.

So we have a precision carriage, with a closed-loop, direct current drive mechanism. All we need now is a means to control it effectively. The sawyer is still the key to successful headrig operation—but we can make his job easier and more efficient by giving him the best tools available.

This is where the computer comes in. Many people will back away at this point with justified concern about the complexity of computer controls and related maintenance. I will not say that a computer is really simple and anyone can service it. It is *not.* It is a complex piece of electronic hardware; but properly selected, applied, programmed, and maintained, it is a valuable tool. A competent electrician can acquire enough know-how in a few days of training to handle basic computer maintenance requirements.

Figure 9.7 illustrates the function of the computer in setworks control. The computer stores all of the information necessary for proper setworks operation. This includes basic motor control data (such as maximum speed, high or low jog speed limits, forward and reverse controls, and the travel limits) as well as the setworks functions, such as set dimensions. The complete setworks then consists of the sawyer's controls (pushbutton and readouts), the computer, the SCR power supply, the direct current motor, and the pulse generator. The closed-loop system is now complete.

Setworks features and functions

The flexibility of the closed-loop system is shown by the functions it can perform. The setworks can be programmed to provide any or all of the following functions:

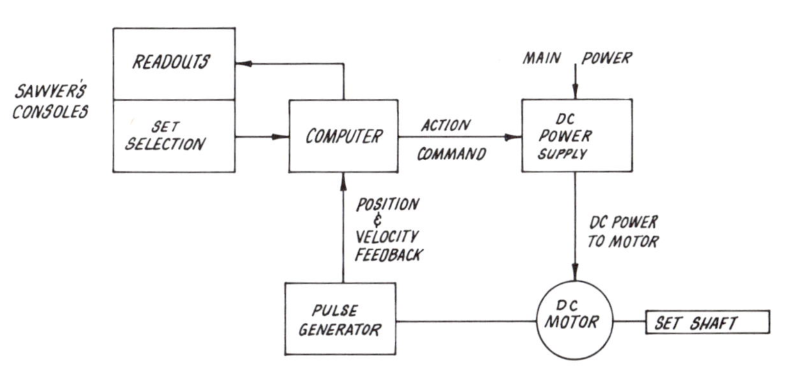

Figure 9.7. Relationship of the computer to other elements in a closed-loop, precision setworks system is shown. Computer stores all information necessary for proper setworks operation.

1. Board or incremental sets.
2. Backstand or absolute sets, also known as cant sets.
3. Set multiples, such as doubles, double-doubles, and triples.
4. Set adders for special products.
5. Backstand memory. The computer will store any backstand distance in memory, with a digital readout for the sawyer, and retrieve it after turning the log.
6. Stacking or run tally. Tally capability permits the use of the set buttons in the same fashion as an adding machine, to create a run from actual set dimensions.
7. Tally to backstand position. The automatic set, out to the tally total dimensions or run start point.
8. Alternate species (or program) sets. By a simple single-number change, all the set dimensions can be converted to a different species combination.
9. Automatic sequence selection can be provided to store preselected run sequences in the computer memory for call-up according to the backstand dimension available.
10. Saw kerf adjustment allows for adjustment of one number to change the saw kerf allowance in all sets (about a 10-second job).
11. Secondary breakdown kerf allowance provides for incorporation of resaw kerfs, or gang or edger kerfs, in the set adjustments themselves for the multiples (such as doubles, triples, and gang cant or module sets).
12. Backstand readout provides digital information to the nearest 0.001 inch of the actual knee position relative to the saw line.
13. Supervisory computer tie-in may be used to connect the carriage/setworks to a log-scanner computer or business systems computer to coordinate carriage setworks with log data, market data, inventory, and so on.
14. Software travel limits. Front and rear limits of

motion, including dogs out and hook dog limits, can also be incorporated.

15. Computer reprogramming can add or convert functions if time and experience dictate the need. This can be done with virtually no downtime and little physical change to the equipment.

Key setworks factors

The possibilities of a precision setworks system are almost overwhelming, but the key factors are these three:

First, all functions are performed accurately and repeatably. There are no cut-and-try adjustments.

Second, the sawyer is relieved of mental arithmetic problems, permitting him to concentrate on grade. Also note that if, at any time, the sawyer notes a change in grade or *any* unusual situation, he can manually interrupt any automatic sequence and proceed according to his judgment.

Third, all pertinent data—including backstand dimension, memory dimension, if used, and the like—are presented in digital form to the sawyer to the nearest 0.001 inch. This sometimes takes a little getting used to for the sawyer—to change his thinking from fractions to decimals—but since he does not have to do the actual arithmetic, the transition should be smooth.

Many features of the setworks system fall into the category of sawyer aids. The sawyer has accurate dimensions of backstands and sets. Kerf allowances are automatically included. The tally, or set stacking feature, eliminates the mental arithmetic and rough estimates of getting on a run. Further, the system eliminates the mental fatigue associated with these functions. The sawyer knows exactly how much material he has to work with and how much of it he will be using.

Comparison of accuracy

The above describes briefly most of the features and functions to look for in a precision carriage and closed-loop setworks. The next question is, "What can be gained by using this type of machine and control system?"

Let us make some comparisons. I have checked setworks repeatability on conventional open-loop systems manufactured by our company, as well as that on other systems. All the systems tested lived up to their claimed accuracies—namely, plus or minus 1/32 inch—approximately 85 percent of the time. In the other 15 percent of the time, accuracies went to plus or minus 1/16 inch and in some cases as high as plus 3/32 to minus 1/16, or vice versa—a total span of 5/32 or 156/1000.

For the most part, these checks were made on new machines with new setworks under better-than-normal operating conditions. Giving full benefit of the doubt to the conventional open-loop system, we can figure a *best case* accuracy of plus or minus 1/32 inch. Conventional open-loop is here considered to be a limit switch, potentiometer balance, or similar control with clutch-brake stopping or a start-stop control on the motor.

Accuracy of a precision carriage with a closed-loop, direct current setworks can be summarized as follows (this is a *worst* case condition, assuming all factors to be at maximum tolerance). The setworks drive and control itself, up to the motor shaft, has an accuracy of plus or minus 0.001 inch. The spiral bevel gear has a maximum backlash of approximately 0.001 inch at the gear or 0.0001 inch at the knee. The ball screw and nut, using a non-preloaded nut, has a backlash maximum of 0.005 inch. The lead error of a commercial rolled-thread ball screw is 0.001 inch per inch maximum. So the total *worst case* accuracy of the precision carriage and setworks combination is plus or minus 0.0071 inch (Figure 9.8). Note that we can reduce the backlash of the ball/nut to approximately 0.001 inch by use of a preloaded nut, and the lead error can be reduced by a factor of 10 to 0.0001 inch by use of a ground-thread ball screw instead of the rolled-thread ball screw.

The plus or minus 0.007 inch *worst* case on the precision machine compares to the 0.031 or 1/32 inch *best* case on the conventional system. The first comment that usually pops up (and it is a legitimate one) is, "So what? If you can't get another board out of the log, what difference does it make?" Well, let us see whether it does make any difference.

First, we can agree that nobody wants thin boards. If we must be off a little in dimension, we should always be a little on the

PRECISION CARRIAGE

WORST – CASE ACCURACY

SETWORKS DRIVE & CONTROL (AT THE MOTOR SHAFT)	.001″
BEVEL GEAR BACKLASH	.0001″
BALL SCREW / NUT BACKLASH	.005″
BALL SCREW LEAD ERROR	.001″
TOTAL	.0071″

Figure 9.8. Elements determining "worst-case" accuracy which might be expected with a precision carriage and setworks combination.

heavy side. We can always take the excess off, but we cannot thicken up a scant board. This irrefutable condition tells us how we must apply our tolerances.

Let us use an 8/4 board (Figure 9.9) as a reference thickness—having a finished thickness of 1.5 inches; a shrinkage allowance of 1/16, or 0.062; and a finishing allowance of 1/16, or 0.062; or a total undried, unfinished target dimension of 1.624 inches.

Since we do not want a thin board, we must change our setworks' tolerances to read for the conventional setworks plus 1/16 minus nothing, or plus 0.062 minus nothing. For the precision setworks we will have plus 0.014 minus nothing. These allowances put all possible error on the heavy side and say that we must add half of the total tolerance in each case to our target size. Thus, the conventional target dimension goes from 1.624 to 1.655 inches and the precision target dimension from 1.624 to 1.631 inches.

Now add a factor for knee-to-knee misalignment: 0.005 inch for the precision and about 0.031 (1/32 inch) for the conventional. The net difference in *target* dimension is then 0.050 inch. This means that for every 8/4 board cut, we have to allow at least

ACCURACY COMPARISON
(CONSERVATIVE)

BEST CASE CONVENTIONAL	.0310″
WORST CASE PRECISION	.0071″

TYPICAL 8/4 BOARD

	CONVENTIONAL	PRECISION
FINISHED DIMENSION	1.500″	1.500″
SHRINKAGE (1/16″)	.062	.062
FINISH (1/16″)	.062	.062
TARGET DIMENSION	1.624″	1.624″
CARRIAGE / SETWORKS ALLOWANCE	.031	.007
TOTAL TARGET DIMENSION	1.655″	1.631″
KNEE-TO-KNEE MIS-ALLIGNMENT	.031	.005
GRAND TOTAL TARGET DIMENSION	1.686″	1.636″

Figure 9.9. In a "conservative" accuracy comparison of open-loop and closed-loop setworks performance, "best-case" accuracy of conventional is compared with "worst-case" accuracy of precision system.

0.050 in extra thickness to provide for carriage and carriage setworks inaccuracy. It does not seem like much—but remember, we used the *best* case conditions for the conventional system and the *worst* case conditions for the precision system.

The extra 0.050 inch represents about 3.1 percent extra material allowance on every 8/4 board. Again, a very small number. It will not produce an extra board in every log, but it will reduce the volume of material removed at the planer. It will, with accurate run start points, almost eliminate shims and it will upgrade the overall material yield from the log. The larger the log, the better the yield improvement. And remember, we are using a conservative number, that is, 0.050 inch on every board. This is about the *least* improvement that can be expected. A more realistic number would be much greater.

After a little wear and tear on the conventional drive train, an accuracy of plus or minus 3/64 would be a more likely figure, making our set allowance 0.047. A realistic precision carriage accuracy allowance would be no greater than 0.005 (Figure 9.10).

A realistic evaluation of differences in the two systems

Thus we have a new set of real-life target sizes: for the conventional, 1.702 inches; for the precision, 1.634 inches. Now, the target size difference is 0.068—a reasonable number in my estimation. This gives us a percentage difference on *each* 8/4 board of 4.2 percent.

This still does not assure an extra board from every log—but a 4.2 percent improvement in yield is a very significant figure. It will

ACCURACY COMPARISON
REALISTIC

TYPICAL 8/4 BOARD

	CONVENTIONAL	PRECISION
BASIC TARGET DIMENSION	1.624″	1.624″
SETWORKS/CARRIAGE ACCURACY	.047	.005
KNEE ALLIGNMENT	.031	.005
TOTAL	1.702″	1.634″
DIFFERENCE	.068″ or 4.2%	

Figure 9.10. A more realistic comparison of the conventional and precision setworks systems than that shown in Figure 9.9.

produce one or more extra boards on large logs, and it will improve yield from smaller logs by putting more fiber into boards and chips and less into planer shavings.

In addition, bear in mind that the conventional system, even with good maintenance, is still subject to severe deterioration and change due to factors such as clutch and brake wear, relay closing time variations, and "morning sickness"—to name a few.

The precision system (with no brake or clutch, no chains and sprockets, no relay times to consider, and with the anti-friction ball screw-and-nut combination) is not susceptible to early deterioration and does not require continuous adjustment to maintain accuracy.

Weighing costs—precision versus conventional

The precision carriage dc setworks system is not cheap. In fact, size for size, the combination may cost two or three times its conventional rack-and-pinion carriage and open-loop setworks counterpart. Is it worth it? I think a little history can best answer that question.

In the spring of 1973, at the Southern Forest Products Association Exposition in Atlanta, we were asked how anyone could justify spending so much money for a carriage. It happened that a man from Louisiana, who had recently put one of these systems into service, was standing nearby, so we referred the question to him. His response went something like this: "Well, we did some estimating and based our decision to purchase on a *potential* increase in overrun of about 4 percent. This was to result in a payoff time of about one year. After running for some three months, we found an *actual* increase in overrun of about 10 percent and we then reestimated the payoff in a total of about six months." (More recent data from this source indicate an overrun increase of 12 percent.)

The U.S. Forest Service, in its program of sawmill performance evaluations, recently surveyed the same mill and found a total variation within boards of 0.024 inch and a total variation among boards of 0.023 inch. This was total board measure and considered not only the carriage and setworks but also the bandsaw.

It seems appropriate now to emphasize that the carriage itself is only one of a number of major elements contributing to possible inaccuracy at the headrig. Other elements include the bandmill, the saw itself, the rail, and the structure supporting all the equipment. All these factors should be reviewed critically, and improvements made where needed to be sure that maximum yield improvement is achieved.

Reliability of precision equipment components

Here are some other questions that are usually raised about the precision carriage system: How reliable is it? Do the ball screws and nuts and gears hold up? How often does the control system foul up?

First, the mechanical components: The ball screws and nuts are rated by the manufacturers with a maximum 20-year life under the loads imposed. We estimate, under typical sawmill conditions, a more conservative life of at least 10 years. This 10-year life span should be without significant degrading of accuracy and efficiency. The gearing, with matched spiral bevel gears, will wear somewhat faster—but the gears are shim-adjustable to maintain minimum backlash as wear does occur.

Second, the electrical/electronic components: A heavy-duty, high-response, mill-type direct current motor is a rugged machine. SCR power supplies have come of age and are used in all types of rough industrial applications from steel mills to plywood plants. Minicomputers also have passed the test and are in common use throughout industry.

Initial problems resolved

Have there been problems? Certainly. Any really new system goes through a debugging stage. There were ball screw coupling problems, pulse generator sensor problems, computer program bugs, and wiring and noise pick-up problems—to name a few. But with patience on the part of users, and perseverance on the part of manufacturers, these problems were solved. Systems installed today are starting up and running smoothly with a minimum of downtime.

Weyerhaeuser Co. purchased its first precision carriage dc set-

Top: The Filer & Stowell computer-controlled, precision ball screw carriage installed at Weyerhaeuser Co.'s North Bend, Oregon, mill. Log is just entering carriage and has not yet been dogged. *Bottom*: A cant on the Weyerhaeuser North Bend carriage as seen from the log infeed deck. The 10-foot, double-cut bandmill is sawing largely old-growth Douglas fir timber.

works system and installed it at North Bend, Oregon, early in 1973. Weyerhaeuser ordered two systems but held up delivery of the second until they could evaluate results from the first. The first machine had been in operation about three months when the second was released—it started up in February 1974. Now both are in operation and, I believe, performing up to expectations.

To give an idea of the smoothness of startup on these systems today, we can use the second North Bend carriage as an example. This is a five-block, 84-inch-opening machine just over 44 feet in length. The mechanical installation alone was a big job. Once installation was completed, it took one weekend to check out everything. The first production shift had about three hours of downtime for miscellaneous adjustments and minor modifications. Since then, operation has been virtually trouble-free.

Weyerhaeuser now has four such systems in operation in Oregon, two at North Bend and two at its Cottage Grove mill. A fifth system is scheduled for a mill in South Carolina. Are they paying off? I think the history speaks for itself.

The purpose of development and the reason for use of a precision carriage dc setworks system lie in these two factors:

1. Simplification of mechanical design—to minimize the number of friction high-wear parts by substitution of antifriction low-wear parts with minimum backlash and slack.
2. Provision of a setworks with maximum practical accuracy and repeatability, along with the flexibility of providing the sawyer with accurate readout data and calculator and memory functions.

The net effect of the precision setworks system is a step toward maximizing the product and the value yield from our forest resources. Thus, we maximize our own profit while at the same time conserving those valuable natural resources.

10

Water gun conversion for headrig carriages

Jim Bigelow, President, Bigelow Machine Co. Inc., Portland, Oregon

This chapter will discuss energy-saving equipment, or to put it more accurately, a way to use energy now being wasted in the mill. This equipment is a water shotgun. Installed on a headrig carriage as a conversion from a steam gun, a water shotgun will save up to 10,000 pounds of steam per hour on a 10- to 12-inch gun. The steam which has been saved can then be put to a better use as energy in dry kilns or turbines, for instance, rather than shooting it into the atmosphere as waste. Gemaco water shotguns are available in all sizes, as complete new guns or as conversions for present steam guns.

Conversion without downtime

Conversions from steam to water can be installed over a *long* weekend, with no downtime involved. Mills that have converted to water report even faster carriage response than with steam, sometimes picking up a line or two a minute over steam. Another good time to change to the water shotgun system is when the steam gun needs reboring or reringing. (You will not need to do this after conversion to the water gun—another plus.) And if the steam gun plug is badly worn, we replace it with a fiber plug without rings.

A Gemaco water gun feed installed on a headrig carriage at the Pacific Crown Lumber Co. mill, Plummer, Idaho.

Advantages of steam over water

Of course, sawyers like water because it is much easier to control than steam. Water is positive. It is not delayed like steam. Water will not compress like steam, so reaction down the line is immediate. The water is recirculated into a 1,000- to 1,500-gallon tank. The tank is filled with a 16-to-1 soluble oil solution. This combination helps prevent freezing. The electric pump power is used only while the carriage is in motion. At other times the water is recirculated with a very low amperage draw on the pump motor. Water temperature buildup is handled either by a fan or a water-cooled heat exchanger.

The valve is the key to the water gun

The heart of the gun is a patented valve which is controlled by a hydraulic slave valve. This, in turn, is positively hooked into the present lever of the saw.

Some water guns are installed with the valve outside the tank–a feature which makes maintenance easier. The gun is very rugged. A direct-drive motor operates the pump.

Cooling equipment is necessary

A cooler is needed for this equipment, as the gun will build up heat as it is used. The faster it is used, the more the heat increases. Many guns have a fan cooler, but a water aftercooler can be furnished if the customer desires.

11

Engineering for increased production

Donald E. Hand, Vice President, Dyna-Tech Corp., Portland, Oregon

Edward E. Washburn, Jr., E. E. Washburn & Associates, Portland, Oregon

Part one by Donald E. Hand

Historically, a mill manager would remark that he was in the business of making lumber. Today, we all recognize we are in the business of making wood fiber yield the best economic return. The name of the game is recovery. We no longer simply cut logs—we cut dollars. The underlying theme throughout Sawmill Clinic sessions has been greater yield from each log.

We have heard about recovery from bark, recovery from waste, computerized controls, and scanners. However, I firmly believe that we have not fully utilized our most important asset—namely, our dedicated, hardworking employees.

Consider the man!

We spend literally months analyzing the log mix, material flows, and breakdown machinery. Certainly these factors are critical. However, it seems that when we come to consideration of machine operations, we frequently lose sight of the importance of the interface from the man to the machinery.

When the United States began to think in terms of putting a

man into space, anthropologists gathered data on the shape of man and determined the conditions under which he could work most efficiently. These charts are available to us. Our tax dollars paid for them. Let us use them in the design of human-engineered control systems.

First, let us consider an operator standing at a console. We observe that he has a comfortable field of vision limited to approximately 120 degrees. Next, we see that a console top has an optimal elevation of 42 inches with a 30-degree slope, or slope to 48 inches maximum. Does your equipment measure up to these standards?

Next, look at the same operator sitting down. Here we find that chair height should be 17 inches—but how many times have you seen a man sitting on a stool at a console that was really made for stand-up operation? How many times have you seen a man sitting on a seat fashioned from scrap lumber and a cast-off pad? How many times have you seen pushbuttons located at a height other than the 28 inches which anthropologists tell us is the optimum? How many times have you seen controls located beyond the optimum 30-inch reach of the operator's arm? These dimensions are just a few of the parameters that should be considered in locating and choosing equipment for efficient operation.

Think about location of his controls

Location of each button, switch, pilot light, and similar device should be considered with respect to the machine function, with respect to frequency of operation, and with respect to optimum positioning for the operator. Control stations should be designed with a similarity of location of functions and actions, to permit easy transition by the operator from one station to another.

Further, we should not overlook the man's environment. Take into consideration temperature, humidity, ventilation, lighting, noise, and vibration. Perhaps the Occupational Safety and Health Act, in this area, is calling our attention to a need which has existed for some time. If an operator has a feeling of general well-being, he has the ability and desire to do a better job all day long. The result will be increased productivity and increased yield.

The factors I have described add up to a totally human-engineered control station. Certainly this approach is not inexpensive. The initial expense will be greater than if some pushbuttons simply are located in a steel box. But, when one considers that the equipment will be used for more than 4,000 hours per year, how important is the additional cost? More important, what is the return on this investment? Any measurable increase in productivity or yield will more than amortize the increased cost of engineered controls and continue the return year after year—without additional expense.

With a logical, human-engineered control scheme, a machine operator will be able to direct his full attention to the passage of material and to the functioning of the machinery, rather than to looking for a button to poke. This should bring about measurable improvement in performance of both man and machine.

Production increases?

To my knowledge, there has been no definitive study made in the lumber industry of production increases resulting from this human-engineered approach. However, adequate documentation from other industries points to the wisdom of taking positive action for improvement. Why is a desk 30 inches high? Why is a secretary provided with a desk for her typewriter and papers? The answer is obvious—for more efficient and more productive work all day long. The tough question is this: Why should mill people be expected to be miserable working in a cold, drafty location? Buttons and levers are frequently located for ease of a one-time installation, rather than for ease of operation shift after shift, year after year.

The majority of our employees are serious and dedicated to doing a good job. However, we frequently throw stumbling blocks in their way. Each operator can, and will, do a better job if he is provided with a better work station.

Emphasis must be on reliability

You provide better machinery—consider providing better controls. And with regard to better controls, have the control systems engineered for reliable, continuing operation.

Operator control booths, partly manufactured, on the factory floor at Dyna-Tech's Portland, Oregon, plant. Buttons most frequently pushed are within easiest reach. Cabs are sound, dust, and temperature controlled; and double safety glass is tilted for best view of work. Sawyer's stick was modeled after helicopter stick. Sawyers, like logs, come in various sizes, so chair raises, tilts, and moves backward and forward.

Our industry has emerged from one of sporadic, one-shift operation to, in many cases, around-the-clock operation. The pressure therefore must be for the design and selection of equipment which will operate continuously and reliably. Regardless of care and attention to details, equipment will fail; yet these failures can be kept to an absolute minimum within the constraints of good economic decisions.

Recognizing that failures will occur, equipment design should incorporate features to ease maintenance.

Electrical and electronic technicians in today's mill operation are faced with the repair of all manner of equipment ranging from the powerful, simple headrig saw to the sophisticated computerized scanner and setworks.

How can we expect the mill technician to deal at one moment with a multihundred-horsepower, brute force motor controller and at the next moment deal with what is, to him, a complex electronics logic circuit?

The answer lies in careful design for ease of maintenance. Consider what downtime is worth. Frankly, none of us likes to think about downtime. Let us think positively. Let us think in terms of 100 percent "up" time. This goal may not be attainable; however, with proper planning we can come close to a perfect equipment operating record.

Electronic equipment improvements

Many operators have had a bad experience with electronic equipment, or know of someone who has had such an experience. But do not forget, we are an emerging industry. We are in a period of development. Improvements have been made and will continue to be made. The aircraft industry, as an example, has experienced rapid strides in technology. Starting first with the DC-3, which had only simple communication radios, today fully 25 percent of the multimillion-dollar cost of a 747 is devoted to electronics. The industry's average availability, or "up" time, exceeds 99.6 percent. This fact points out that complex electronics can be interfaced to machinery with increased reliability. The forest products industry too can look forward to this same level of reliability—if not today, certainly in the very near future.

Locating the trouble source quickly

A production manager really does not care what is wrong with the equipment when the mill is not producing. All he cares about is repairing the failure and putting the equipment into operation as quickly as possible.

How can a technician bring sawmill machinery back on stream quickly? He can do it by having state indication lights within the equipment which will indicate the area of failure. These pilot lights should be considered routine in the development of a good control system. These state lights (or lack of light) will indicate location of the trouble. The technician will have a key to the offending part or circuit card and will be able to replace it quickly.

Really, who cares what was wrong? Fix the equipment. Put the machinery back into operation–and then when the dust settles, a technician can analyze the failure and repair or replace the offending part. Meanwhile, production continues.

Electronics technicians are needed

Let us look a little deeper at this technician. You must consider an electronics technician on your staff today because of the advent of more sophisticated equipment. In the sophisticated mill, the day of the combination millwright-electrician has passed. Equipment advancements in the lumber industry require a man who has the ability to analyze and understand electronic failures.

Military spin-offs

Our tax dollars have been spent to train many fine technicians in the military service. We should consider hiring these men to work in mills. Mill electricians can train a technician in areas of industrial electrical equipment, while the technician reciprocates with training in electronics. The technician is familiar with the buzz words such as L.E.D.(light-emitting diode), open circuit transistor collector, optical coupler, and I.C. (integrated circuit) Chip. He can assimilate information about new equipment rapidly and take advantage of the opportunities to learn more about the equipment when it is installed. He should not wait until a failure occurs to acquaint himself fully with the intricacies of the system.

Furthermore, any responsible supplier will ask each mill manager at the time of equipment start-up to allow his men some time with the installation technician. This hands-on training time, which may range from a few hours to several days, can pay handsome dividends later.

We certainly never like to raise the specter of a failure, but regardless of how well a piece of equipment is designed, failures will occur. It is not at all unusual for the unknown electronics to be blamed for a failure when, in fact, the problem might have been caused by a simple limit switch. If a mill technician does not understand how the system is to function, if he does not take the time to learn, if he is not provided with the opportunity and incentive to learn, he may undoubtedly jump to erroneous conclusions. It is basic human nature to blame that which one does not understand. I am sure that we all have had experiences which underscore the need for the understanding of a control scheme.

My favorite experience regards an irate mill manager who called and was rightfully upset that a particular conveyor continued to run when it should not. I was fortunate to have had considerable experience with the control system and the machinery as it was installed in his plant. As I recalled, the tail pulley of the offending conveyor was on the same shaft as the drive pulley of the previous conveyor. (I think the reader probably is getting ahead of me and has immediately guessed what the problem was.) I asked the manager whether he would please have one of his men check the lubrication of the idling tail pulley.

Simple, you say? Obvious? Not really. I had the advantage of knowing the system so well that it appeared to be a virtual impossibility that any failure in the electronic system could have caused the action he described.

Ask for help

When my mill manager friend called back a short time later to advise that a grease gun had fixed the electronics problems, he also said what I had been wanting to hear for several months. "Don, let's get your technician together with our people and spend some time going over the system." Many years have passed and that equipment is still operating very well. Why? Because his technicians understand the operation of the machinery and the electronics equipment. You should expect to receive this type of assistance from suppliers.

The control stations and electrical-electronic control systems in a plant should be selected with the same care and attention to detail as is given any other major item. Typically, machines are supplied with control panels which are not necessarily compatible with each other. Rather than confront an operator with a hodgepodge of control boxes, an integrated control station which will make it possible for one operator to manage an area of a facility should be considered.

I have pointed out several factors to consider in the design of a better control system. I hope that the information will aid mill owners and managers in creating a human-engineered environment for operators and technicians–important members of the team–who are striving to attain the goal of higher yields, greater recovery, and improved profitability.

Part two by Edward E. Washburn, Jr.

Before discussing computerization in a very small log mill, we need to define a very small log. In the past, any log under 8 inches on the small end was considered too small to process. This size was lowered to 6 inches, and now is as low as 4 inches in some areas. In Montana, any log under a nominal 6-inch diameter is not scaled and is considered slash material. Therefore, a mill cutting logs 6 to about 10 inches in diameter should be considered a small-log mill.

This may be a good general description of a *small* log mill, but at Del Conner Lumber Co. our definition of a very small log mill is one that will process material 8 feet long with a minimum diameter of 2½ inches to a maximum diameter of 9 inches. Optimum large-end diameter is 6½ inches.

When the idea was conceived to utilize this material, it was realized that a method of sorting and controlling it would have to be developed. Test runs with visual sorting showed a sort diameter spread of 1 to 2 inches. Therefore, it was decided some kind of automatic scanner system had to be used. By using a chipping headrig system in conjunction with a scanner and minicomputer,

Size of typical logs handled by the very small log mill installed at Del Conner Lumber Co., Darby, Montana.

we found we could get the best lumber recovery while maximizing production.

The problem then arose as to how to set the chipping headrig for maximum production. Feeding the machines at 170 feet per minute and setting for every log, with a set time of two seconds, the maximum number of 8-foot 3-inch blocks per minute would amount to only one block every five seconds, or twelve 8-foot 3-inch blocks per minute. Whereas, running the 8-foot 3-inch blocks butt to butt, one size at a time, would give us 20.6 blocks per minute—a 42 percent increase.

So it was decided to sort the logs before they got to the chipping headrig. Thus we could run one size at a time. With only one size of lumber coming out at a time, the lumber could be stacked automatically. Production was then based on one 8-foot 3-inch log every 10 feet of feed speed—or 17 logs per minute, which equals 1,020 logs per hour.

Equipment choices

With this concept, equipment which would do the job had to be located. Design of the electrical equipment, scanner, and computer was turned over to Dyna-Tech Corp. The chipping headrig job went to Adco West. The debarker is a Finnish Model VK 16 Valon Kone from Mac-Fab. It will bark logs from 2 to 14 inches in diameter with a minimum length of four feet. This allows us both to buck the treelength logs before debarking and also to buy 8-foot blocks as cord wood from local farmers.

It was decided to use five holding bins with enough capacity for eight hours' run of any given size and also to switch from bin to bin as required. This gave us five different cutting programs (shown in Figure 11.1):

Holding bin arrangement at the Del Conner mill in Montana's Bitterroot Valley, where small lodgepole pine is a large part of the available timber.

Program 1: Logs 2½-inch minimum diameter to 3 1/8-inch maximum diameter yielding one 2x2 (1¾ x 1¾)

Program 2: Logs 3½-inch minimum diameter to 3 7/8-inch maximum diameter yielding one 2x3 (1¾ x 2¾)

Program 3: Logs 3 7/8-inch minimum diameter to 4¾-inch maximum diameter yielding one 3x3 (2 13/16 x 2¾)

Program 4: Logs 4¾-inch minimum diameter to 6 1/8-inch maximum diameter yielding one 4x4 (3¾ x 3¾) *or* two 2x4s (1¾ x 3¾)

Program 5: Logs 6 1/8-inch minimum diameter and larger yielding three 2x4s (1¾ x 3¾)

By running the logs butt to butt, we increase productivity. We have added a factor for downtime equal to about 10 feet of chain run for each 8-foot 3-inch block delivered to the mill. Using these programs separately and our production capacity of 1,020 logs per hour, production per eight-hour shift is:

Program 1: 22,032 board feet of 2x2
8.96 units of chips

Program 2: 32,640 board feet of 2x3
14.64 units of chips

Program 3: 48,960 board feet of 3x3
15.44 units of chips

Program 4: 86,496 board feet of 2x4
29.76 units of chips

Program 5: 129,744 board feet of 2x4
32.24 units of chips

The above figures give an average production of 63,974 board feet of lumber per eight hours plus 20.2 units of chips—well exceeding the 50,000 board feet per shift that was desired in the initial concept.

Without the use of computers and scanners, this could never have been done. But the efficiency that can be achieved with an automated system makes it possible. Use of a computer system

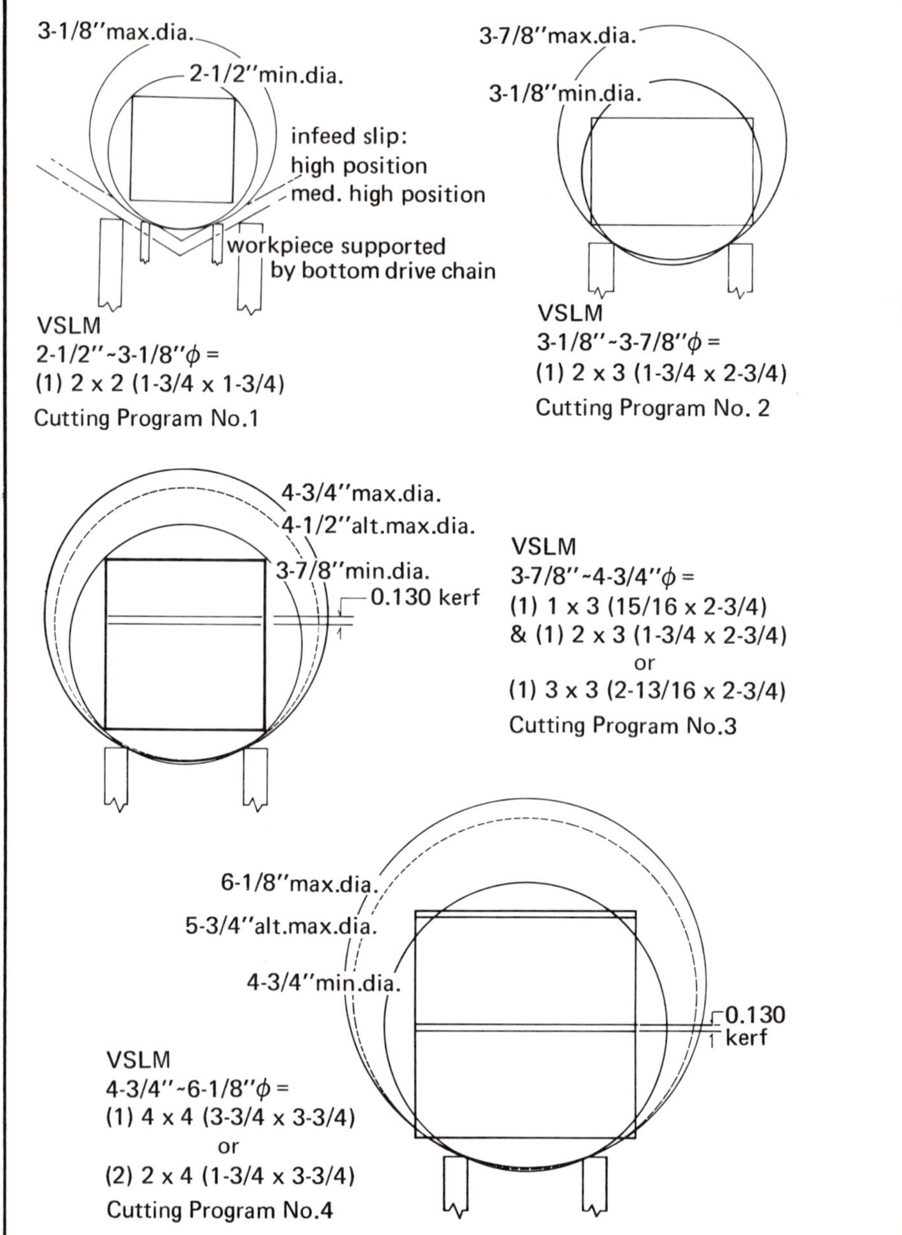

Figure 11.1. Five cutting programs—and lumber recovered—used by the

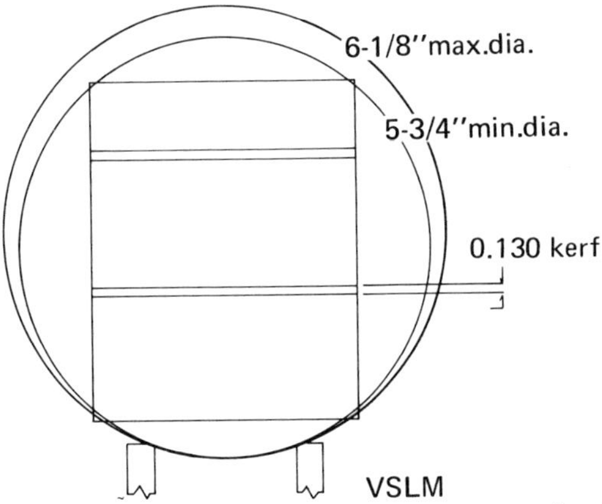

VSLM
5-3/4″~6-1/8″ϕ = (1) 1 x 4 (15/16 x 3-3/4)
(2) 2 x 4 (1-3/4 x 3-3/4)
Alternate Cutting Program No.4-5

VSLM
6-1/8″~7-1/2″ϕ = (3) 2 x 4 (1-3/4 x 3-3/4)
Cutting Program No.5

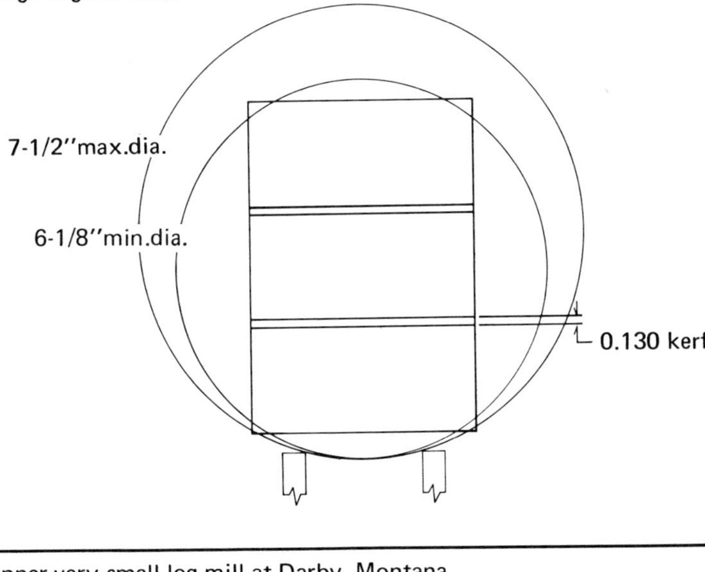

Del Conner very small log mill at Darby, Montana.

also gives a printed readout at the end of every shift showing everything that has been run on the shift.

Almost everything run through this plant is waste that previously was left in the woods as slash.

Mill flow described

The flow pattern of the mill begins at the log infeed deck. Logs are placed on the 35-foot infeed deck in bundles by a LeTourneau log handler and broken down by two sets of reversible deck chains. Logs are then fed one at a time onto the 10-foot slasher infeed deck to the buck-up saws. We have a John Deere hydraulic loader remotely operated from the operator's control booth to straighten any logs as they go onto this deck.

Logs move through a six-saw slasher arrangement with ending rolls. The log comes onto the ending rolls. The operator decides which end of the log is the butt end and, depending on this decision, sends it either to the left or to the right so the butt may be cut off. The waste material or the small part of the log is always dropped off the top. We do not measure logs in the woods. We bring them into the mill and buck them at 2½ inches.

After bucking, logs are fed into a 52-foot infeed conveyor to the Valon Kone debarker. Feed rate is 197 feet per minute. Logs are kicked off the slasher saw table automatically, keeping the

Left: A 6½-inch-diameter log on the infeed conveyor, ready to enter the debarker. *Right*: Log leaving the debarker. Effort must be made to keep log ends butted for maximum production. Debarker feeds at 197 feet per minute.

conveyor ahead of the debarker full at all times. So we are running continuously through the debarker.

A brief comment about chipping. Bark on lodgepole pine is quite thin. We are taking waste material from lodgepole, before it is debarked, and dropping it onto a conveyor to the chipper, where it is incorporated with our other chips. We are still staying under the 1 percent maximum bark content in our chips. We expect that the small chipper that handles wastewood from the existing Beaver mill will also take trim blocks and lily pads that we are chipping for the paper mills.

Electronic information-gathering system

After debarking, the 8-foot 3-inch blocks travel through the Dyna-Tech scanning system. The information gathered is fed into a minicomputer (housed in a clean, pressurized environment) for storage. The scanners and minicomputer tell us everything we want to know about the operation. At the beginning of each shift, we feed the cutting program we will run into the computer. As the log travels down the conveyor (at a speed of 240 feet per minute), the scanner picks it up and feeds the information to the minicomputer, which decides which of five holding bins the log will be kicked into.

Storage allows high-speed processing

Storage, or holding, bins are 45 feet long. We stack logs to a depth of 6 feet, which gives quite a bit of capacity. Stacking into the bins requires that the logs be kept straight. We have a chute down each side. Without the chute we found that some logs could travel 20 or 30 feet after we kicked them off sideways. We kick the logs into a set of arms tensioned by a cylinder at the top. The log progresses to the bottom of the bin, slowed by the arms. The bin is adjusted so that the next log that comes off will kick the arms out, the arms will catch that log and drop the bottom log into the bin. The accumulator arms accumulate a stack before laying over so we have a 6-foot stack height. We tried kicking logs into the bin with a man's foot. At the speeds we were running, it was physically impossible to keep up without electronic control.

Log in chute at lower left is about to enter the scanner. Log size is sensed, and a memory system kicks each log into the appropriate bin beyond outfeed chain in background. Bin conveyor travels 240 feet per minute.

From the storage bins, logs are broken down in a trough system. A jack ladder takes the logs one, two, or three at a time and transfers them onto the infeed conveyor to the Adco West VSLM (very small log mill) chipping headrig. We have a holding area on top where the logs can spill over and are spread out into a stop-and-loader. The stop-and-loader then kicks them on. Photocells tell us when the conveyor is empty.

We try to maintain about a 6- to 8-inch gap between logs on the chipping headrig infeed conveyor. Logs are fed automatically into the chipping headrig. We have a roaming operator in the area to make sure everything is fed through smoothly, to take care of any jamups, and to maintain enough stickers in the sticker stacker to

Two views of the holding bins, before construction was completed. Holding arms are stanchions tensioned by a cylinder.

Jack ladder carries logs from holding bins to conveyor entering the mill. Logs are accumulated in saddle at bottom of photograph; logs move up jack ladder on arms, as shown.

stack the material. The headrig is, of course, set to match the cutting program for the log size stored in the bin being run. We try to maintain same-size output—either all 2x2s, all 2x4s, or whatever it is we are cutting. If the market demands, we can pull our 1-inch boards off to the side, but we would prefer not to do this.

Same-size output allows automatic stacking

Lumber output from the chipping headrig then travels onto an infeed chain to the automatic sticker stacker, which stacks the loads to be fed into the dry kiln. We have a three-load storage capacity. When the stacker is full, it kicks the load out. If the forklift driver has time, he puts a band around the stack. When we began handling this material, we decided to use a lumber carrier. But we found that, with the large logs (three 2x4s) a lumber carrier would have had to make a round trip from the stacker to the dry kilns in six minutes to be ready to pick up the next unit. It took the lumber carrier seven minutes to make the trip, so we were stuck. Going to multiple storage allowed the forklift to take two units at a time—a twelve-minute round trip to the dry kiln and back to the stacker.

Same-size lumber output eliminates the need for a green chain and allows the stacker to stack all the mill's production.

Automation and the scanner/computer control system allows the mill to be operated by three men—a log deck and debarker operator; a chipping headrig and stacker operator; and a roving millwright.

Using this mill in conjunction with the existing bandmill and Beaver operation achieves complete utilization of the available raw material—and, to reiterate, the material this mill processes previously was considered waste material.

DISCUSSION PERIOD

DICK SETTERSTROM, State of Montana, Butte, Montana: Would you comment, Ed, on the dead timber you have been receiving during the past year in the Darby area?

WASHBURN: We have been running dead timber through the mill, but this is out of my field. I can tell you that we have had good recovery. We have run some of it through the Beaver. We do have some in the yard at present. It is feasible to process this wood. We are logging dead logs that have been on the ground and we are processing them. It is working.

QUESTION: Would you explain your bucking operation?

WASHBURN: The buck-up rig is nothing more than a glorified trimmer. We have six saws on the deck. Logs travel through the saws at a feed speed of 80 feet per minute. We drop them into a cradle and hold them, then send them through. On a 40-foot log we can get five 8-foot blocks through this deck system. This is the one reason we went to the slasher system. We looked at chop saws but found we could not buck fast enough with them. The slasher saws will stay ahead on anything that is 16 feet or longer.

12

Measuring efficiency at mill stations

Wayne Bennett, Manager, Lumber Research & Development, Weyerhaeuser Company, Tacoma, Washington

We will start with objectives. I once asked one of our business managers what his objective was. He said, "Two words–make money." Then he looked me right in the eye and said, "No, three words–make money quickly." The overall objective of the industry obviously is profitability. But there is a more specific objective.

I want to examine the many factors of sawmilling that affect profitability, particularly around the cutting stations or machine centers. The subject is yield and grade. If we all understand these factors and how to measure them, then we will make progress toward improvement.

A sawmill is a system. It is an extremely complex system, especially if operated optimally. The overall system is made up of a series of subsystems that we call cutting stations or machine centers. Each subsystem must be operated in an optimal manner, and the strategy for the total mill must be correct, to gain maximum efficiency. But what is maximum efficiency?

In engineering terminology, *efficiency* normally means output divided by input, resulting in a number that is less than one. The meaning of high efficiency here is high profit. If output value of the product is divided by input costs and the number is less than one, the sawmill operator is in big trouble.

There are two major functions in attaining high efficiency. One is making the correct decisions; the second is executing those decisions.

To make the best decisions we must have information about (1) raw materials, (2) the capabilities of our converting machinery, and (3) the product lines we are trying to make. We will concern ourselves with conversion capabilities.

The conversion diagram shown in Figure 12.1 starts at the left with logs input. At machine center No. 1, the operator takes out a cant, which becomes the raw material for machine center No. 2, and so on through the sequence. Chips and low-value residuals are dropped out in the process.

Conversion capability is determined by the factors shown in Figure 12.2. System restraints, item 7, include the size of the pieces, the number of sorting pockets, or whatever limits are built into the capital facility.

Accuracy, smoothness, and kerf

One mill objective is to minimize the sum of three items: kerf, standard deviation of accuracy, and roughness. Of course, the operator may have to increase kerf to get accuracy under control, but what he is really looking for is the minimum sum of the three, not necessarily the minimum of any one of the factors.

We are talking about the minimum use of wood per saw line

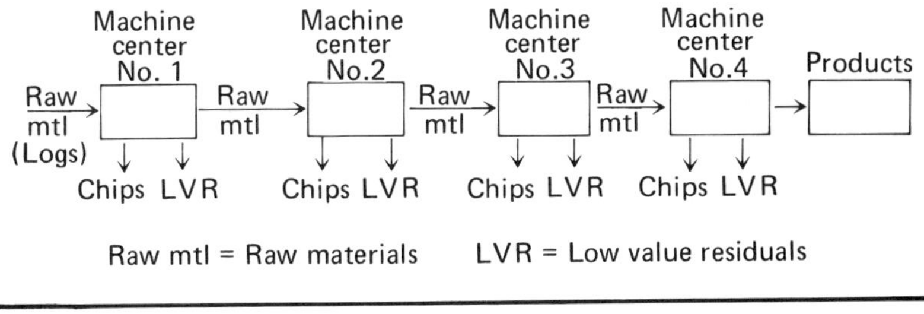

Figure 12.1. Raw material conversion process.

Raw material → Conversion → Products

Converting capability determined by:

1. Ability to start cutting where desired
2. Ability to cut in the direction desired
3. Ability to cut accurately in chosen direction
4. Ability to cut with smooth surfaces
5. Ability to remove minimum kerf
6. Cost of attaining above
7. System restraints

Figure 12.2. Factors determining conversion capability in the mill.

cut. A histogram of accuracy is the statistical means of showing variation in sizes. A size two standard deviations negative from the mean size is generally used as the minimum size that will dress out in the planer. Total wood use per saw line, then, is the average piece cut, plus the actual kerf taken (Figure 12.3). And the actual kerf is usually larger than the actual tooth width of the saw.

It is not correct to use ranges in talking about accuracy. One piece of wood that is either much too small or too large can double the range. Range really does not accurately indicate what is occurring. The only number that means anything statistically is the standard deviation. Figure 12.3 also shows that if the range of size variation were reduced and the operator wanted to reduce the target size to take advantage of that reduction, he could only reduce the target size by half the range reduction.

Which aspects of a cutting station affect efficiency?

Now let us look at those aspects of a machine center or cutting station that affect efficiency (Figure 12.4). First we must consider ability to position the piece with respect to the cutting tool. Then we must evaluate the transport accuracy—how straight can that piece be conveyed past the cutting tool? Third, we must examine the alignment of the machine and machine components—that is, the alignment of arbor, tracks, and transport system.

Machine vibration is a severe problem under impact loads in a sawmill. Also involved is vibrational stability of the saws themselves, the problem of saw filing. This includes precision in sharp-

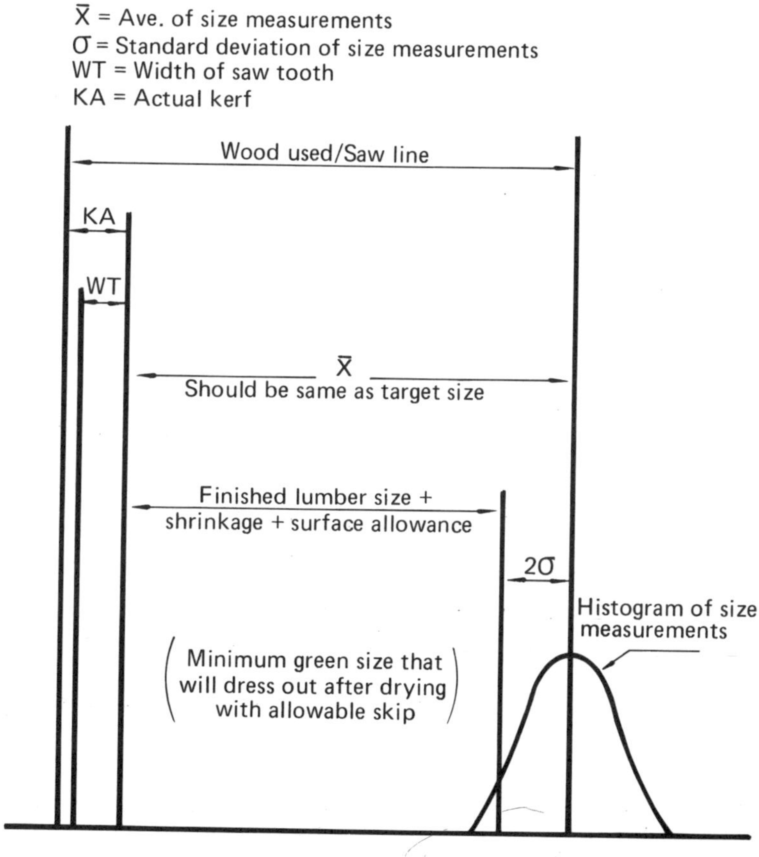

Figure 12.3. Wood use per saw line equals average piece plus actual kerf, which is usually larger than actual tooth width of saw.

1. Positioning accuracy
2. Transport accuracy
3. Machine alignment
4. Machine vibration
5. Saws and guides
   a. Vibrational stability
   b. Saw filing—precision and sharpness
   c. Saw geometry—teeth and plate
   d. Guide type and alignment
6. Power requirements
7. Setworks
8. Operator
   a. Raw material alignment
   b. Feed speed
   c. Performance observation (maintenance)

Figure 12.4. Aspects of a machine center that affect efficiency.

ness and symmetry and it also involves saw geometry—tooth angle, gullet size, break angles, side angles, clearance, and so on. Also to be considered are the guide type and the alignment of those guides.

Then there are power requirements. Adequate power is required to keep saw speeds up to the point where control of vibrational characteristics is possible. Setworks precision has been discussed widely and many ingenious devices have been developed.

And finally, in speaking about parts of a machine center that affect efficiency, we must not forget the operator—a very real part of the operation. In most cases, he has the job of aligning material with respect to the saw. In some cases he determines the speed at which material moves through the saw. That speed often has a profound effect on the accuracy, on the roughness, and on the actual kerf taken. Of course, a very real part of his job is to watch performance so that he can call for a saw change or let the foreman know when there is machine trouble.

Why measurement is important

Now let us consider testing and measurement, particularly the importance of valid measurement. The following statement was attributed to Lord Kelvin many years ago: "When you can measure what you are speaking about, and express it in numbers, you know something about it; but when you cannot measure it, and when you cannot express it in numbers, your knowledge is of a meager and unsatisfactory kind." Progress demands that we be able to measure what we are doing and to express it in statistically valid numbers.

Statistical measurement allows the manager to make rational decisions, rather than subjective, irrational ones. It allows the manager to determine what incentives may exist for making a suggested change or taking action. It allows the manager to monitor the performance of an operation so that he will know when to call for maintenance or to make a change. Statistical measurement allows the manager to determine the effect of change. For example, if you buy a new rig and put it in the mill, how do you know whether it is really paying off unless you measure its performance in terms of profit? How does a manager know what the return on his investment is? Usually he simply guesses. How many managers actually make a full report on a capital investment so that they *know* whether they got the return on investment anticipated? Statistical measurement allows improved maintenance that really can optimize the operation and maximize profit.

We must measure both decision-making and execution. Is the man who makes decisions about each piece of stock going through a machine making the correct decisions? I know only two ways of determining whether the correct choices are being made.

Methods of checking decisions

One method of measuring operator decisions is to put a skilled observer beside that man to observe and tally what he is doing. Such an observer can watch what a man is doing at a trim saw or at an edger and tally whether or not he is doing the right thing. The separate observer has the time. He need not worry about handling the piece and pushing it through. He has time to observe. The real problem with the observer technique is its "halo" effect. The operator knows he is under observation, so his performance does not necessarily represent long-term performance.

The other technique for measuring operator decisions is photographic. The piece can be photographed before and after processing at a cutting station. An advantage of this technique is that the camera can be permanently mounted above, turned off for a month or so, then turned on by remote control. The operator does not know it is on, so the halo effect is eliminated. The before and after photographs can then be compared.

This comparison is very important. We often measure the products that come out of a machine in terms of accuracy, roughness, kerf, and so on. But we rarely pay much attention to how much the operator threw down the hole. Once wood hits the chipper, there is no way to put it back together. The relationship between what enters and leaves a station is called the fiber balance. It is very important that the right decisions be made so that good pieces do not get thrown down that hole. And there are only two techniques we know to help make those decisions—the skilled observer and photography. (Videotape would qualify as a modified photographic technique.)

Methods of checking execution

Next we turn to measurement of the execution process. We are talking about the actual kerf, the standard deviation of accuracy, and the roughness of the surface.

Measuring actual kerf taken

Let us consider actual kerf. It is very difficult to measure actual kerf taken on a saw. The tooth of the saw can be measured; but that, for many reasons, is not necessarily the actual kerf taken. First, no filer can level a saw closer than 0.005 inch. Most saws are leveled to 0.016 or 0.020 inch. Secondly, saw vibration affects kerf. Unless tension and saw heating are kept under control, there is sure to be some vibration. This vibration makes a saw act like a dado, which produces a kerf somewhat larger than actual tooth width.

Here are some techniques used to measure kerf. Put a square-sided, pre-weighed cant through the saw, shake off the sawdust, and weigh the pieces. By mathematics (assuming some average density through the piece), one can arrive at an approximation of kerf taken. Another method is to take a squaresided cant, measure the width, take the sum of the statistical means of the individual pieces cut, divide that by the number of saw lines taken, and thus arrive at an actual kerf. There is yet another technique, which is quite unsatisfactory. This technique involves cutting part-way into the piece, then stopping. The problem with this last method is that as the saw slows down, it goes through critical speeds and starts dadoing. So the kerf measured does not necessarily represent what actually happens.

Now the results: on edgers we find in many of our saws that the actual kerf taken is between 0.030 and 0.100 larger than the width of the tooth. That represents a very substantial amount of money. We worry about getting one or two mils of correction in the setting. We talk about cutting 10 mils off the tooth width. Yet we are often cutting kerfs 0.030 to 0.100 inch larger than the tooth.

Measuring standard deviation

Now let us talk about how to measure standard deviation. The classic way to do so is using technicians with calipers. I think by now all our operations are so converted that we can talk in terms of thousandths of an inch. It is very important that we measure in decimals and not in fractions. There is a great need for instrumentation that can measure standard deviation, to eliminate the cost of having technicians do it. Weyerhaeuser and others have research projects underway to monitor thickness and roughness automatically for each piece behind each machine center. Thus we can monitor performance.

Measurement of roughness difficult

Measurement of roughness is usually quite difficult. It is a very sophisticated subject in the machine tool industry, involving root mean squares and a number of other rather sophisticated techniques. We have chosen to talk rather simply in terms of measuring the depth of the grooves between the peaks and the valleys to indicate roughness of the surface. To do this we have developed a very simple field device. It has two parallel steel ways on which precision round-way ball bearings are mounted. There is a small

A technician with calipers measures lumber width for accuracy of cut.

Lucite carriage that runs on the two steel rods. Placed on the Lucite carriage is a dial indicator with a pointed stem projecting out of the bottom of the groove in the bottom of a channel-shaped aluminum housing. We place the housing on a piece of lumber and run the carriage back and forth. The dial indicator has tattletales on it, which give the maximum and minimum motion of the stem, thus indicating roughness. It is not exact, but it gives a fairly good indication of which saws are going bad and what kinds of saws perform best in terms of minimizing roughness. Roughness is a very real part of wood loss.

There are other ways of measuring roughness. We have precision profilometers with needle-fine points in the laboratory that can do all the sophisticated work. But those are not practical for a mill environment. However, we have found our field device to be quite practical. We have also made some simpler versions that can be fitted into a pocket that do a quite creditable job.

Now let us go back and talk about measurement at all the machine center parts. I have discussed how to measure the decision process and the execution process. Did the man decide correctly what to do? Did he really do it? Suppose you conclude that the process is out of control, the products are not right, or that the operator is making errors. We have discussed all the parts of the machine center. But how do you measure what specific parts are doing?

Evaluating execution—positioning the piece

The first place where measurement is needed is in positioning. We talk much about such things as best opening face and scanning and computer programs. We have used these for many years in our company. But one of the real problems is that we have no real means of executing so the sawyer will know where that best opening face is. We say he should cut 6 inches (or whatever) from center. But where is center? It is something he judges roughly on a log. But to get high yield out of small logs it is extremely important that the sawyer be precise to about one-tenth of an inch as to where he opens that "best opening face." Much development work needs to be done on how to position logs and cants with respect to the cutting tool. We have all the logic, we have all the strategy, we have the electronics and the scanning—but we do not have the hardware. I know of only one mill that has that capability—Sun Studs, Inc., Fred Sohn's mill, at Roseburg, Oregon. He does have that capability. That mill was individually designed and developed by Fred Sohn and his consultants.

Transport through the cutting station

Let us move on to transport—the ability to convey a piece of wood past a cutting tool. Let me tell you about some of the troubles we have had. We had evidence from research that, if a fence was not used, roll feeds were bad actors in terms of cutting a straight line. So we did a good deal of research to try to find out why. We found that the trouble had nothing to do with the squareness of the cant and it had nothing to do with the sawing pattern on the surface. It really had to do with the basic charac-

To check lateral movement of cant through saw, a straight line is made on cant face by stretching wire, then overpainting and removing wire.

teristics of the wood itself—its variable compressive modulus. So we started observing what was happening at some of our round-saw machines.

We set out to test the ability of a set of roll feeds to feed through an edger in a straight line. The rolls were aligned carefully; new bearings were put in. The first thing we did was to mark a straight line down some cants. We did this simply by stretching a wire and then overspraying it with paint, leaving a white line in the middle of the paint field.

Next we moved the saws to the side and put the cant through the machine at full running speed, probably around 250 feet per minute. We taped a yardstick to the machine frame just below the guide holders. As the cant is run underneath the roller, the white line (as displayed in the paint field) will shift back and forth with any lateral movement of the cant.

Cant moves through saw under yardstick to measure lateral movement as revealed by the unpainted white line.

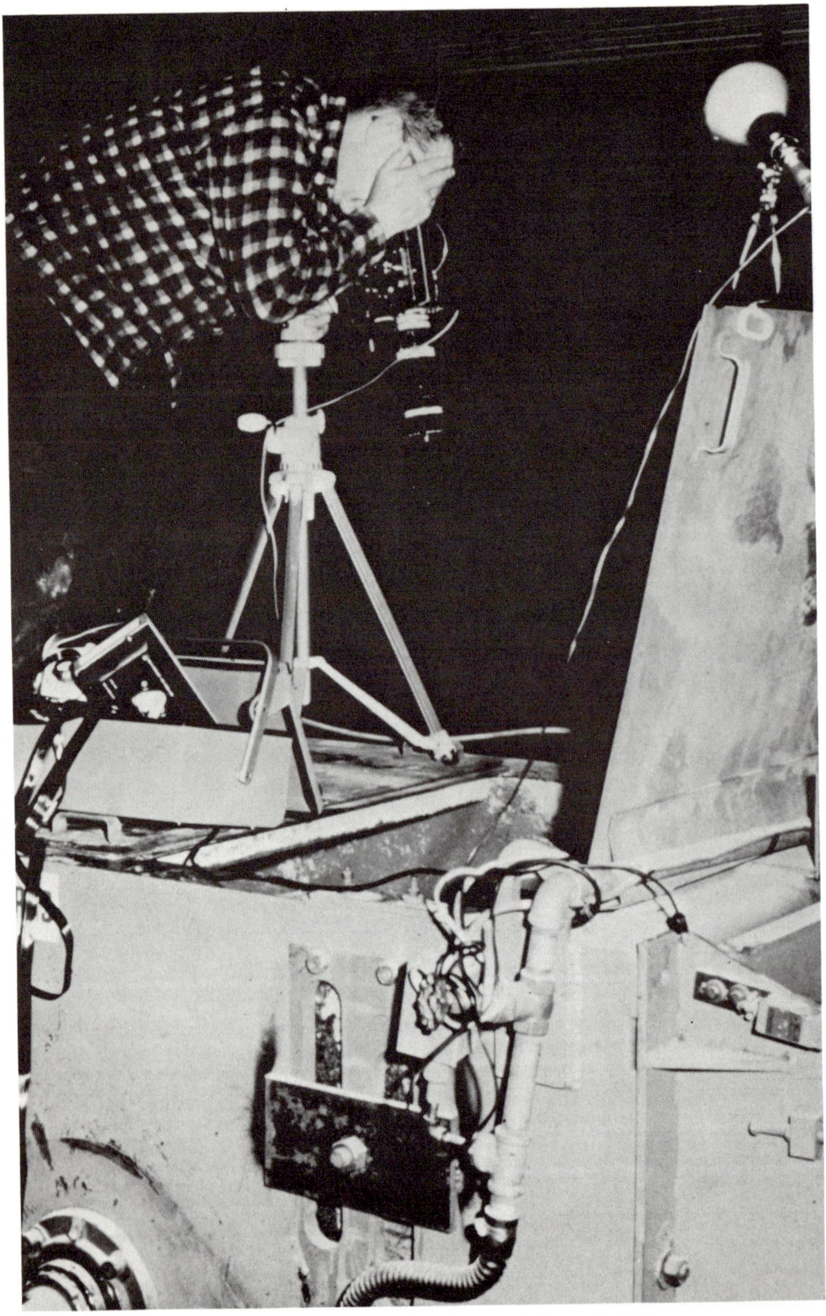

High-speed camera above test edger records cant's lateral shift.

As the cant went through the machine, we used a high-speed camera mounted on top to take a picture (slides, so that they could be easily blown up) every one inch of travel. We analyzed those slides in the laboratory to see what happened to the deviation—the lateral shift of that cant with respect to the machine frame. There were no saws. The machine was running at full speed without any influence of saws. The cant was carefully aligned at 90 degrees to the arbor before starting.

Here is what happened. In some cases there was as much as 1¼ inches in lateral shift in just 16 feet of travel. In one case, we found we had a 3/8-inch lateral shift in only 2 inches of forward travel. Can you imagine what that is doing to the saws? We have been busy trying to figure out how to cure this problem. We have made some progress but we are still working.

Our answer so far is to convey the piece of wood through a machine on a precision track, such as a carriage track. The best way we could find to do that was to build a high-speed slat bed feed. This feed was designed for 350 feet a minute. This feed happens to be on our research edger at Longview, Washington. (The edger is in the sawmill in production but is a research machine.) We built the slat bed feed and had some problems debugging it. We have installed a number of these feeds in our sawmills. We still have not licked the positioning problem. But once we get the cant locked in that machine on the Coleman points, on top of the slat bars and under the pressure rolls, it travels in a straight line. Rolls are urethane coated to avoid any lateral pressure on the cant as it goes through the machine.

Minimizing machine vibration

Now let us talk about machine vibration. It is vital that the vibration of the sawing machines and the surrounding structure be under control. Saws have natural frequencies of vibration that can be self-excited under certain conditions. There are also resonant frequencies at which they will start vibrating based upon excitation from either the machine itself or from an adjacent machine. It is important to get the vibration under control. Saw vibration is a contributor to excess kerf, and profit loss.

We in Research and Development render the technical service to

A technician testing vibration on the test edger.

our Weyerhaeuser Co. operations to balance band wheels, chippers, couplings, motors, arbors, and all other motion-critical parts. Several of our larger complexes now have their own equipment and trained technicians.

Improving accuracy of machine alignment

Let us move on to machine alignment. Presently the technique for alignment is the taut wire. It is certainly not accurate. Weyerhaeuser and others are working on optical laser systems for cross-alignment of bandmills, alignment of tracks, alignment of guide tracks, and alignment of arbors. We have developed a number of laser devices for aligning our edgers. In our opinion, alignment must be done under dynamic conditions–under full load and speed–not under static conditions with a taut wire.

We have a precision test mandrel in our laboratory. Many of our production saws that were giving trouble with dadoing are tested here. With instrumentation, we can monitor precisely what the saws are doing. We can monitor vibration and determine plate thickness, collar diameter, or tension to get away from the critical speeds of that saw. The precision mandrel has less than 0.0001 inch runout in the bearings, so we measure vibration of the saw only. With test equipment of this type, we can get the whole system under control and know what we are doing. The readout comes on a strip chart. The narrow line on the chart is an index of the runout of the saw. This is a measure of flatness of the saw ("level," in filer's jargon).

As we increase saw speed, we can pass through critical speeds where the saw starts to vibrate. The line on the strip chart becomes wider, indicating a dado effect. Poorly designed saws can vibrate or flutter as much as ¼ inch. So it is vital that saw vibration be controlled.

Many people are researching the problem of saw vibration. Among the foremost is California Cedar Products. Others include the University of California's Department of Mechanical Engineering and the U.S. Forest Products Laboratory at Madison, Wisconsin. Weyerhaeuser and others are funding research on this problem.

Precision laboratory test mandrel for testing circular saw vibration. Vibration is recorded on the strip chart at right.

The saw—filing techniques, tension, and tooth geometry

Now let us discuss saw filing. Precision and sharpness in the grinding room are very important. Management must provide precision equipment, tools, and training for saw filers. Much work needs to be done in the field of tension, particularly on how to measure it. Some research is being done. Among the displays at the Sawmill Clinic was one by Letson & Burpee showing a new technique for measuring the so-called tension of a bandsaw.

Another field that needs work is tooth geometry. Our literature is full of tooth shapes. I once talked to a manufacturer of filing-room equipment who had documented various tooth shapes. He had stacks and stacks of drawings of different tooth shapes. I asked how many he had for bandsaws and he said he had about 1,500. But he was quick to point out that only 75 are in common use. Well, when 75 different kinds of teeth are used to saw wood, that says one of two things—we do not know what we are doing, or none of it makes any difference anyway. Much research is needed on the correct tooth geometry for the job at hand.

Guides and guide alignment are very important. There is a great need for quick-setting fixtures to be put inside sawing machines so the filer or the millwright (or whoever is doing the guide alignment) can do it quickly and precisely. It is a great deal of trouble to get guides in exactly the right position.

Precision and repeatability necessary in setworks

We also need precision and repeatability in setworks. Let me cite a case. At one time we were developing a digital setworks for a carriage. It was very precise. With a dial indicator set on the front of the knees, we could set within 0.002 inch, time after time after time. So we thought, "Great, now we'll put wood through to see what happens to sawing accuracy." You can guess what happened—there was no improvement. The rest of the sawing system was out of control. Either the tracks were not aligned or the saws were vibrating or something else was happening.

Figure 12.5's chart shows (at the top) the actual kerf, the standard deviation, and the roughness. In the left-hand column are the various parts of the machine that influence those factors. The X's show what parts of the machine affect the sawing variables

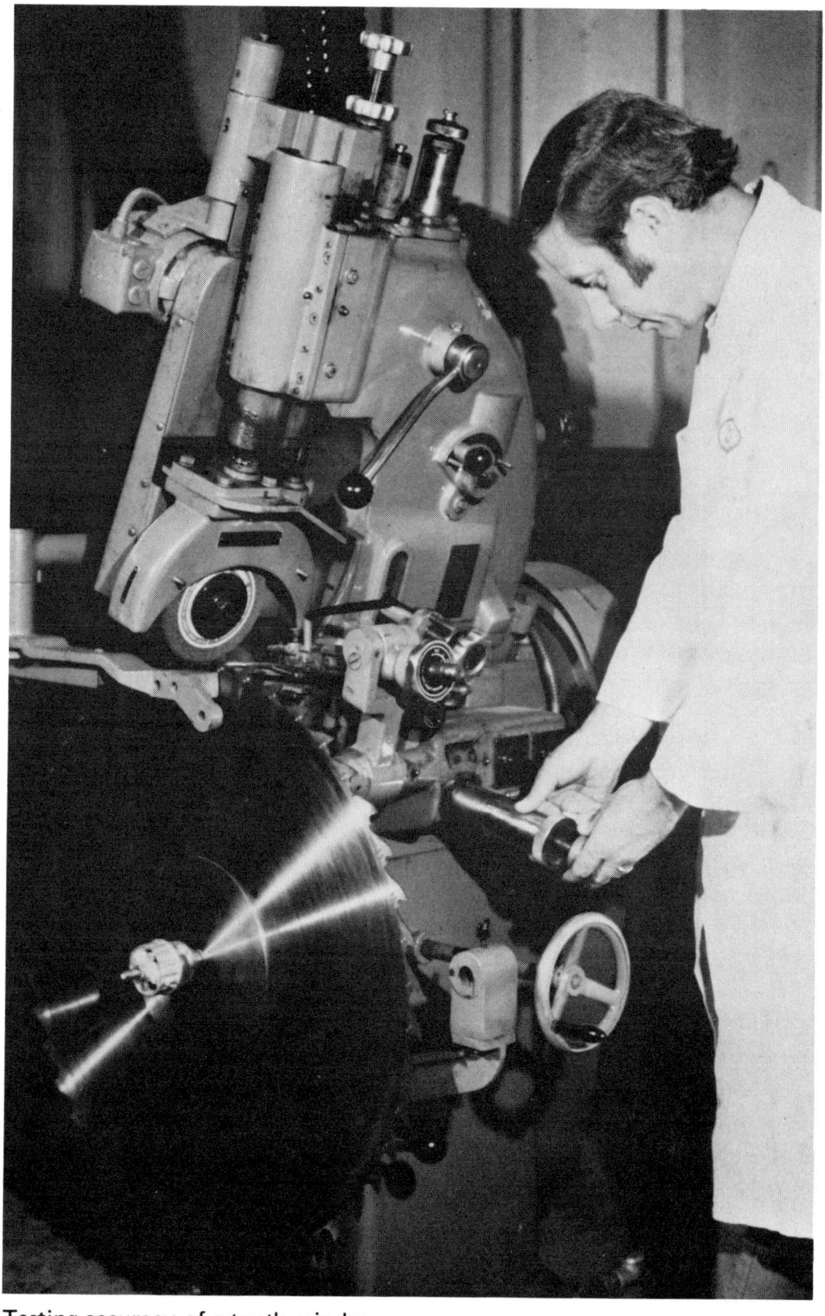

Testing accuracy of a tooth grinder.

Adjusting saw guides for absolute accuracy is one of the most important jobs in a mill. Quick-setting fixtures are needed to aid filers or millwrights in doing a quicker, more precise job.

being measured. There is an extremely complex set of interactions among these variables.

Better means of measurement are needed

One of our great needs is measurement—we need instrumentation. We can get much of this measurement by hiring enough technicians to go out and take these measurements. But that is quite expensive. We need on-line instrumentation.

Three kinds of instrumentation are needed. First, we need instruments to make measurements for use in decision-making. How wide, how big, how thick, how long? Where are the checks? Where are the knots? We need these measurements to help us make the right decisions about where to cut, how fast to cut, and whatever else we are doing.

Second, we need instruments to monitor cutting station performance. If we monitor the performance of one machine center,

	Actual kerf	Standard deviation of accuracy	Roughness
Transport	X	X	
Machine alignments	X	X	X
Machine vibration	X		X
Saw stability	X	X	X
Saw filing	X	X	X
Saw geometry	X		
Guides	X	X	
Power	X		X
Setworks		X	
Operator		X	X

Figure 12.5. Parts of a machine affecting three sawing variables—actual kerf, standard deviation of accuracy, and roughness.

use that information correctly, and can keep track of that piece of wood, that monitoring can provide decision-making information for the next operation. We need to know the size and shape of products from a cutting station. We need to know their roughness.

Finally we need instruments for maintenance. We need excellent instruments for machine alignment, for guides, for transports, for positioners, for bandmills. We need instrumentation for saw preparation to measure runout and precision. How many mill operators, in their filing rooms, have a simple dial indicator and a mandrel for mounting saws to check the runout of the saw just by cranking it around in a circle? A *great* need is a means of measuring tension quantitatively. Tooth symmetry—is one tooth right behind the other? Sharpness—how sharp are the teeth? Saw vibration measurement should be in the filing room. Then there are tooth shape, gullet, angles, and all the rest.

We have run several tests with the Portland Iron test edger. Figure 12.6 shows the statistical data we gathered. The standard deviation of accuracy is the vertical, the Y axis. Along the bottom, on the X axis, are the refits of the saw. We started with brand-new precision-ground saws at zero time. Then we ran them. The refit

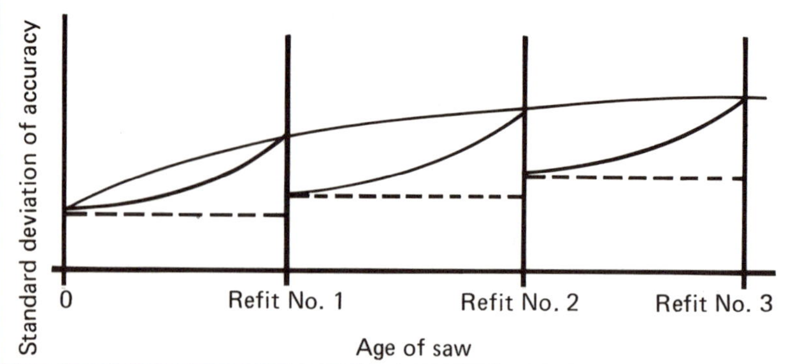

Figure 12.6. Decay curve of saw accuracy.

numbers (1, 2, and 3) each represent cutting about 2,600 6-inch cants. As we cut, we measured the standard deviation of accuracy of the lumber.

We all know that as a saw dulls, its accuracy drops. At zero time with a brand-new saw, we got an excellent standard deviation. It was about 0.009 inch. As the saw dulled, the accuracy decreased, as shown by increased standard deviation. After 2,600 cants the saw was taken off and resharpened. One might have expected it to reachieve its original accuracy, but it did not. Each time the saw was used, cutting became less and less accurate. If we draw a curve connecting the accuracies at the end of each refit, we have the accuracy decay curve. The question is, when to do the maintenance? When should that saw have been taken off—at the end of four shifts, or some other time? That is the kind of information that can be collected if you do take statistical numbers in your operation. And with such numbers, changes can be made that improve profit.

A sawmill with its machine centers is very complex. Improvements will come as we understand the economic impact of the variables as they constantly change. The decay curve of accuracy is one of the changes. We must base operating and maintenance decisions on economic logic, not unsubstantiated opinion.

13

Rube Goldberg gimmicks smooth mill flow

Melvin R. Paulus, Manager, Lebanon Machine Co., Lebanon, Oregon

B. T. "Burt" Randall, Sawmill Consultant, Superior Plywood Machinery Co., Portland, Oregon

Larry L. Webb, Vice President, Webco Lumber Inc., Grants Pass, Oregon

Richard L. DeArmond, Vice President, Idaho Forest Industries, Inc., Coeur d'Alene, Idaho

Part one by Melvin R. Paulus

Since three of the contributors to this chapter are too young to be familiar with Rube Goldberg firsthand, we turned to Vern White for assistance. We found out that Rube Goldberg was a cartoonist who dreamed up extremely complicated, impractical plans—requiring immense effort—for accomplishing relatively simple, straightforward tasks. When I discovered this, I hardly knew whether the invitations had been indictments against the companies represented on the panel or against their representatives as individuals! But it soon became apparent that the "Rube Goldberg" label was tongue-in-cheek. The subject was "getting the most done with the least effort—and coping with the complications involved."

However, it occurred to me that, as we continue adding to existing plants to increase production, we often create something approaching a Rube Goldberg cartoon: a mass of machinery and conveyors running, it seems, into corners—a maze from which it is difficult—or practically impossible—to find a way out.

It often appears that the only solution to such a situation is to start over. But this chapter will approach the problem from the viewpoint of one who must take what he has and make the best of it. The ideas we will present may not be the ultimate in technology, but they have worked for us.

An important aspect of the "Rube Goldberg" ideas we will present is that many of them were the brainstorms of the men in the mill. When his ideas are encouraged and given a fair chance, the average Joe Blow in the sawmill can be one of the best friends management can have.

If handled properly, these ideas can produce a sense of accomplishment, be used as part of an incentive program, and stimulate the crew to increased productivity. Each of us knows the sense of accomplishment he has felt in watching one of his ideas—whether it is one piece of equipment or an entire sawmill—start to produce results. No paycheck can match that sense of satisfaction or that feeling of accomplishment.

Figure 13.1. The double-arbor scragg headrig at Lebanon Lumber Co.'s Lebanon, Oregon, mill cuts peeler cores. Scragg was designed by a Lebanon millwright and built by Lebanon Machine personnel.

High-production modification

Here is an example of an idea generated in the mill. One of our millwrights designed a four-saw scragg with all the saws on one arbor. This machine doubled Lebanon Lumber's production capacity within two months. Previously, we had been running with an American headrig and an American carriage with hand set. After putting in the scragg, production jumped from roughly 25,000 to 30,000 board feet per shift to 80,000 feet per shift. Although the machine would not have won a prize for appearance, it was an idea that worked.

Figure 13.1 shows the double-arbor scragg machine in use today at Lebanon Lumber. The machine is capable of handling 14-inch logs, has 3/16-inch saw kerf, and is a carbide system. We have run up to 30 logs per minute through the machine, and we can easily average 25 with smaller logs. For larger logs, we slow it down.

Rapid log infeed achieved

Although Lebanon Lumber is exclusively a core plant, we do have machines operating on small logs in a number of locations. Figure 13.2 shows the infeed system to a double-arbor scragg. Note that we are trying to emphasize Rube Goldberg ideas. Many of the ideas in this loading system are from the average man in the mill—the millwright or the man operating the machine. A scanner system is installed on this machine. It is completely computerized and can be run totally automatically. However, we do have a bypass system so we can switch circuits and run the machine manually.

Kicking small logs into a machine capable of processing logs at speeds of 25 to 30 per minute can be a major problem. How can we kick logs in fast enough to keep the machine going? Figure 13.2 shows an idea generated in the mill. We use a regular cradle with a set of kicker arms tied to it. When the cradle comes up into a receiving position, the kicker arms go to the "up" position. This offers a greater possibility of fast roll into the feed line. This solution was determined in part by the way in which the mill had been established. It would have required major alterations to change the infeed from the log lift up. Since we did not have a step down that we could use for gravity feed-in, we had to provide

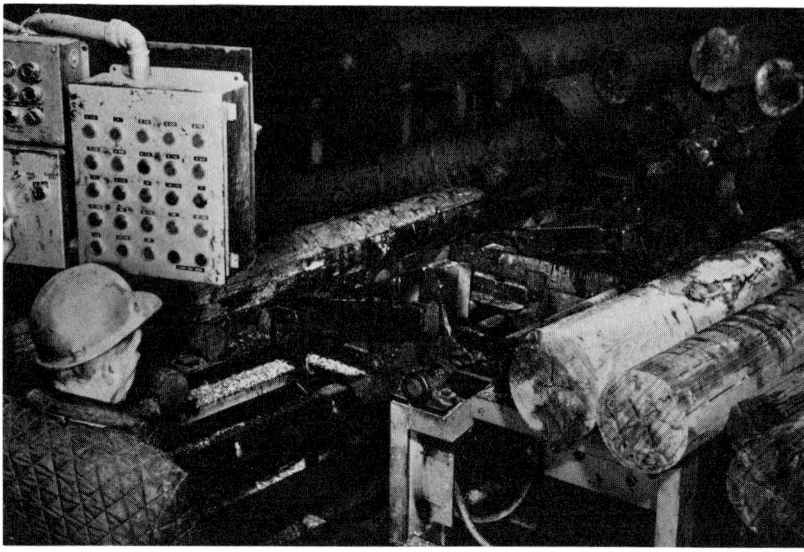

Figure 13.2. The log infeed system at Lebanon Lumber Co.

a mechanical means for rapid infeed. The solution described grew out of that necessity. (Incidentally, we are still working on this problem; we are not completely satisfied with our solution.)

Helping hands at edger infeed

Figure 13.3 shows the infeed to a double-arbor edger. We are feeding at approximately 400 feet per minute to carbide saws. In this particular plant, an edger operator may be required to handle as many as 80 pieces per minute, counting the 2-inch cants and the 4-inch cants. So we have a three-split, triple-roll edger. The top feed rolls are split in three places. The men responsible for this idea, Buck and Wayne Merritt (it was primarily Buck's idea), were at the time considered the average men in the mill. Today they have their own mill, the Merritt Lumber Co., at Priest River, Idaho. Notice in Figure 13.3 that we need helping hands to line up the lumber rapidly for a straight cut. We have what we call skate wheels. The lumber goes under the wheels, which line it up with the saws and feed it through.

Figure 13.4 shows three pieces of lumber going through the

Figure 13.3. Infeed to double-arbor edger provides triple split. Angled rolls throw outside cants against the fence, while center piece is held by skate wheel at rear center of photograph.

Figure 13.4. Three cants are edged simultaneously by a single edgerman and "helping hands" of the Rube Goldberg gimmicks. Cants of various thicknesses can be run in the same pass.

edger simultaneously. One man feeds this machine. Note the 4-inch cant going through on the right-hand side. A 2-inch cant with wane on both sides (center) is lined up by the skate roll to get the lumber rapidly to the linebar for a straight cut. We have added a left-hand pineapple with left-hand spiral on the pineapple. This is what is used to help line the lumber up and put it through the machine.

Lumber separation for smooth flow to trimmer

Figure 13.5 shows another very simple idea originated by someone in the mill. Coming out behind the edger is a cant that has been cut into 2-inch-wide pieces. To break it down for ease of sorting before grading, we use two sprocket wheels (visible in the center and the right foreground of Figure 13.5). The lumber comes up to the sprocket and goes over the top. This simply separates the pieces evenly as they go over the wheels. That is all there is to it. It is a very simple idea, yet it works. And it helps smooth out the flow of lumber to the trimmer.

Variable lumber line needed at planer

At the planer, we had the problem of 12-foot, 8-foot, and 6-foot lumber—all on one line. We needed a variable lumber line. We had a Sweed puller and we wanted it to pull out the odd pieces that we did not want. The planerman came up with the idea of mounting the Sweed puller on a table and putting a crank on it. When he is running 8-foot lumber, he can move the puller into the forward position. For 12-foot, which brings the lumber line back in the back side, he moves the Sweed puller back. Thus he is able to use the same piece of machinery for two applications. This idea is shown in Figure 13.6.

At our plant, we consistently work off a center lumber line, rather than a lumber line off the end of the mill, or end of the lumber. We center all our logs and the finished product with machines, rather than shove everything to one side. We have found a center operating line to be more effective in a stud mill.

Behind the planer we also had a similar problem. We would run 6-foot lumber for perhaps three or four hours. Later, we would change to 12-foot. So we needed some way to change our lumber line behind the planer. The top photograph, Figure 13.7, shows an idea that one of the men developed. He set a roll case on a set of

Figure 13.5. Edged pieces arriving on the sorting line are separated by riding over the sprocket wheel system in center and right foreground.

Figure 13.6. The variable Sweed puller on the trimmer line can be repositioned to handle various lumber lengths.

sprockets. Note the chain along the bottom. Again we simply move the entire unit back and forth behind the planer to adjust the lumber line. Very simple, inexpensive, and it works. Figure 13.7 (bottom) shows the modified roll case in a retarded position, changing the lumber line. Part of the soundproofed planer housing is also visible.

Forklift access necessary

What we have at the planer station is two planers back to back—two lines. We have a central cart system between the two green chains. We wanted to install a roll case for an automatic banding system as an add-on feature. The question was: How will we get in with a forklift? The simple solution we found was to build a roll case with two retractable sections. When the roll case sections are retracted, there is room for the forklift to clear. When the forklift has passed through and there is lumber to be transferred, the roll case is closed. Figure 13.8 shows the roll case in both opened and closed position. Lumber goes over and through a Signode strapper and bander, one of the nicest systems I have seen for automatically banding and end-painting lumber.

Figure 13.9 shows the finished product—what we all aim for.

Flow to trimmer smoothed

Coming back again to some helping hands in smoothing out the flow of lumber on the green chain, Figure 13.10 shows a short stub planer roll mounted in the green chain to line up the lumber with the linebar, getting it ready for the trimmer. We use these systems on the planer chain to move the lumber quickly and very inexpensively. The roll is tapered, with spirals. Some lumber flow systems require a roll made with left-hand, instead of right-hand, spirals. Lumber comes over the top of the tapered spiral roll and then moves to the linebar.

Smoothing out stacking flow

Here is another idea from the men in the plant. Figure 13.11 shows lumber going into the automatic stacker. There are times

Figure 13.7. A roll case set on sprockets can be moved back and forth behind the planer to adjust the lumber line. The photograph at bottom shows the same machine, repositioned to handle a different length.

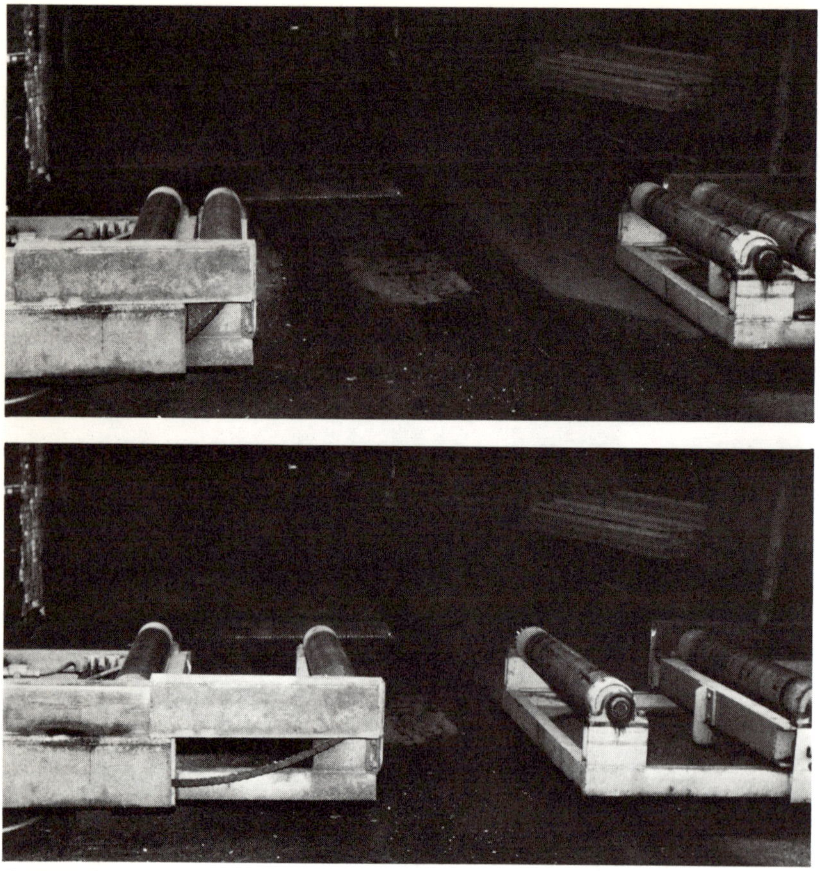

Figure 13.8. *Top*: Green chain roll case in open position to admit forklift. *Bottom*: Roll case closed, allowing lumber to pass down the line.

when the operator wants to hold the lumber back a little to give himself time to straighten out something before it goes into the stacker. Our idea is simply a set of pin stops mounted on an air cylinder. It is a Rube Goldberg idea, but it certainly smooths the flow and makes the operator's job much easier.

An example comes to mind, something I saw in a mill. The edger operator was unable to stop the lumber going into the edger as he wanted. So he got two 2x4s and put them underneath the flow of cants going to the edger with the chain. Then he took another 2x4 and laid it flat. He somehow hooked a cable around

Figure 13.9. A package of finished lumber at the Lebanon plant.

Figure 13.10. This short stub planer roll lines lumber up with the linebar. Tapered spiral roll, set 90 degrees to the lumber flow, is the key.

that 2x4 and then lowered it to the floor. Each time he wanted to stop the lumber, he would raise the two 2x4s. The lumber would come up on it and stop. When I saw this, I couldn't believe my eyes. This was a Rube Goldberg in the true sense, that's for sure. And I doubt they were getting much production, but the edger-

Figure 13.11. A stop point in the stacker line. If the operator needs to hold up the incoming pieces, he actuates the air cylinder pin stops.

man did stop the lumber. And he installed the device himself. Others of you may have seen such "solutions" in the past. We are more sophisticated now, but some of the same principles can help manpower if we give them the opportunity to make suggestions and offer solutions.

I think that is another advantage of some Rube Goldberg ideas. We cry production, but I wonder whether we as management have always been fair in giving the crew the helping hand it needs to increase production. There are many men who could do a better job if we would just give them a helping hand once in a while.

Trimming solution

Figure 13.12 (top) shows two hand trimmers–installed in reverse of their intended position. For this particular application, we

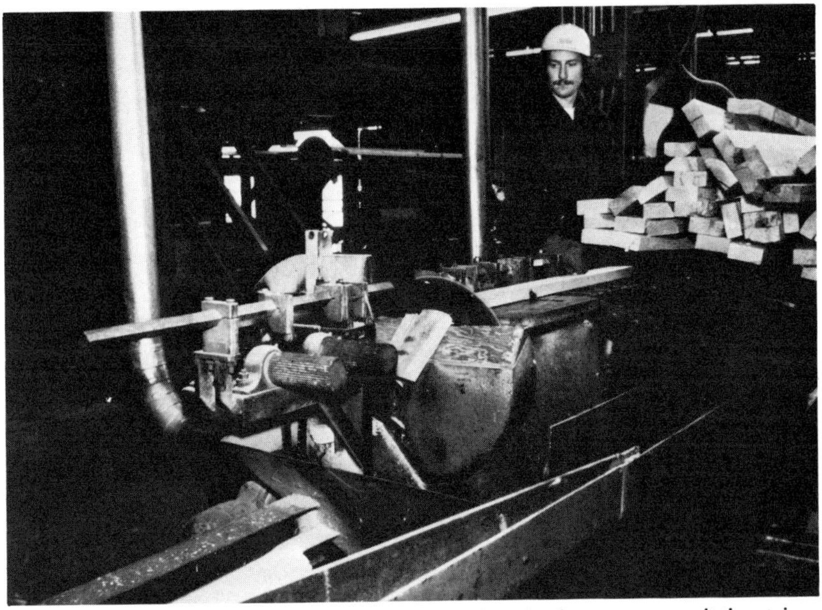

Figure 13.12. *Top*: Two reversed Irvington hand trimmers, permitting trimming from the "wrong" side. *Bottom*: Rolls in foreground rotate clockwise toward the operator, pushing pieces back toward him. Rolls permit trimmerman to bring back anything longer than a 2-foot module. Shorter pieces fall into the chipper line.

needed to be able to bring the lumber from the other side and it was faster for the trimmerman to trim behind the lumber. A hula trim, of course, would have been better, but this was the equipment in the facility at the time so this was what we used. We had the problem of getting certain pieces back. Somebody came up with the idea of using two rolls.

In Figure 13.12 (bottom) the trimmerman is cutting. If the piece is long enough, it moves onto the two rolls shown. Anything shorter, he does not save anyway. The two rolls run slowly toward the operator. He may be select-trimming for odd lengths—54s, 48s, or whatever. If he needs a piece to come back to him, it automatically comes back off the rolls. I would not advise installing such a device in a new mill; but in an existing operation, the idea might help in a particular spot.

Part two by B. T. "Burt" Randall

All operations begin by getting the log into the mill. It has to be bucked. We have found that bucking with electric eyes is an excellent method. The bucking system can be semiautomatic or completely automatic. It can be very quick. The chains can be run very fast and you do not tear out an electric eye when you break the beam. On the other hand, mechanical stops and similar things have to be repaired continually. And above all, bucking with electric eyes is cheap.

This is one of the essentials of a Rube Goldberg—it must be cheap. It has to be designed, not for a mill producing 500,000 board feet a day, but for one that produces perhaps 100,000 to 150,000 board feet. It has to have a payoff time that can be seen in the near future.

For example, there is another gimmick that we have been working on: turning logs end-for-end so as to bring them into the mill small-end first or large-end first, as preferred. To me, there are various advantages in bringing a small log into the mill small-end first. Where the small end approaches the saw first, it is much easier to gauge sizes. Also, most 100,000-foot-a-day operations do not have computers to determine measurements. Turning a log end-for-end can be a relatively simple process and it can be automatic. Just run the log up a small incline on a turn table. Where it breaks over is the log's center of balance. Rotate it and run it off in the other direction and the job is done. This can be done directly in the existing line. It usually is not necessary to tear the mill down to put it in.

The advantages of turning the log can be quite obvious at the headsaw. There is an additional advantage in edging because the shorts can be released. With relatively long lumber, you can release the short and long pieces at the same time. This prevents pile-ups caused by the shorts dropping on the chains and moving down under the longer pieces. There is an added advantage: the narrow end of the cant is at the operator's end where he can see it. It is close to the edger and this will help the operator place his saws into the piece.

Now we have square-edged lumber. If you pick the edging, you have lumber you can do something with: it can be stored. Storing lumber for a 100,000-foot-a-day mill has some obvious advantages. The cutting rate in such a mill will probably be about 125,000 feet. Pauses in production and other slowdowns bring the end output to 100,000 feet. The only way to even out gaps and surges and get a good flow of lumber in the mill is to store it a little. This does not mean storing 20,000 feet of lumber. But a 100,000-foot-a-day mill will produce about 12,000 feet per hour, or around 200 feet a minute. Obviously, if stored (and this is relatively easy to do) 1,000 feet of lumber is five minutes' production; 2,000 feet is 10 minutes' production. Ten minutes' production will allow quite a bit of leeway at the back end of the mill. It also will permit smoothing out surges from the edger and from the headsaw. Storing also permits an even flow of lumber out the tail end of the mill: the sorting chain is not smothered and the trimmer and the resaw are not overloaded.

If the mill is cutting for a resaw and stacking up doubles, assume that the average piece of lumber produced in a day is a 16-foot-long 2x6, or 16 board feet. With 6 inches of chain, 16 board feet can be stored for a limited time. If the piece is a double for the resaw, 32 board feet can be stored in that distance. All that is required is a feeder that can be stopped at will, allowing the lumber to accumulate behind it.

It is not really necessary to break the chain, because once there is a certain amount of pressure built up on the feeder and the desired amount of lumber has been stopped, the next step is to pick the lumber off the chain. This is a relatively easy operation, since standard machinery for it can be built in any sawmill. The process can start over with another feeder. By stopping the lumber in sections, you can store well over 1,000 feet of lumber–1,500 to 2,000 feet of lumber in the average small mill. And it only takes 30 feet of chain to do this. It helps a little if the chain is going uphill–this dilutes the force of gravity on the chain so there is not quite as much pressure.

To even out flow, one requirement is that when you begin feeding the lumber from storage, it must be fed one piece at a time. So you set the feeder to feed just a little faster than the flow from the edger.

Storing also gives a great deal of flexibility. If lumber piles up at the resaw, it slows the resaw operator down. You want to give the operator all he can take–no more, no less.

A storage system such as the one I have described has certain requirements. It must operate unattended; it must be semiautomatic; and it must release lumber one piece at a time. It works on square-edged lumber. Thirty or forty feet of chain is enough–and any old chain will do.

What about monitoring the sizes of lumber at the trimmer? Here you have control of pieces of lumbers on chairs, so it should be possible to monitor these sizes. We are doing some experimental work on this now to see whether this is possible. It should be rapid and cheap. It is not necessary to have a readout in the office nor to compute how many board feet a day the mill is producing. In a small mill, there is usually someone in the office doing that already. There is no need to put in a computer to do this. The amount of money involved is small, perhaps about $5,000, not the $100,000 required for a computer.

By monitoring the lumber, it is possible to know whether the mill is making it on-size all the time. Most mill owners insist on cutting the lumber to the very least size tolerable. When you are allowing only one-eighth of an inch on all sides to plane, you have to hit it just right. And to keep hitting it just right, it is necessary to monitor the sizes. When logs cost a couple hundred dollars a thousand on the stump, it is vital to minimize waste.

Another simple and inexpensive idea is to keep a continuous piece count on all machines. Set it up so there is a readout in the mill superintendent's office that can be seen any time during a shift. You can resaw as many as 20 pieces a minute, and from 15 to 20 would be an average piece count. It is interesting to know how much of the day is spent getting that piece count.

The mill should also know how many licks the carriage hits a minute. How many sawmillers know how many times the carriage goes up and down the track? I would guess, not one in ten. How many sawmillers know how many pieces of lumber come out of the mill in a day? This is important information. It tells you how fast the trimmer must be run. If the average mill, at 100,000 board feet, produces about 3,000 to 5,000 pieces of lumber in a day, that is not much in piece count. So it behooves us to know what is going on at each machine at all times. Then maybe we can juggle the cut around to get another revolution out of the machine or the man. This is not going to cost any more—the man is hired for 8 hours anyhow—so do not worry about that.

Also, if the mill has a twin resaw, the size of the cant going into that resaw should be monitored. There are sizes so close that the operator cannot visually detect the difference between them. One-eighth of an inch difference in cant size might mean two different cutting patterns. Each of these cant sizes are cut exactly to begin with. How is the operator to recognize the extra eighth of an inch from three or four feet away, when there is nothing to compare it with? It is very difficult. Paint spotting is one solution. Another is to monitor the size of the cant before it hits the resaw and display the reading. Again, I am not talking about anything that cannot be bought right off the shelf in a number of places. Monitoring the cant width takes only a counter, an electric eye, and a readout.

The additional lumber volume recovered—although small—will more than pay for the equipment in a few weeks.

Part three by Larry L. Webb

I would like to review the advantages of a quad band edger—the edger we use at Webco. It is the key factor in smooth flow

The Salem quad edger at the Webco, Inc. mill east of Grants Pass, Oregon, runs about 180 board feet per minute and provides extreme versatility including breakdown sawing of defective white fir, which could be expected to break up on a carriage headrig.

throughout the mill, because we run everything through it. We run 16-gauge saws with 1/8-inch kerf. We tried 17-gauge, but Douglas fir is a little too hard. We had trouble with broken teeth, so we switched to a 16-gauge, with which we can still hold the same saw kerf.

We run the edger about 180 board feet per minute. We have three roll cases, all synchronized hydraulically for running wood through. We can go up to 14 or 16 inches high, whatever we want to set the guides at. We can go up to 30 inches wide. Saw change time averages 10 to 15 minutes.

We use this edger a great deal as a small-log mill, too, because we have a slabber in front of our headrig. We take logs 10 inches or smaller, slab them on one side, drop them to the quad, and split them. Once to the resaw for edging finishes the cut.

Recently we experimented running cull white fir logs. As you know, cull white fir is nothing but a "culvert." There are about

A cant comes off the 7-foot Salem single-cut headrig, which cuts essentially for grade. A C.M. & E. slabber reduces number of lines run.

The C & D setworks at Webco controls all four of the 5-ton bandmills at the quad edger to an accuracy finer than 1/32 inch.

two inches of rind on the outside. One cut through the log with the headrig usually leaves a big chunk with no straight edge, square edge, flat surface, or anything else. With a normal edger, you would have to stop everything and pull this big slab out of the mill, or find some way to dispose of it. With this quad, it just runs through. We split it up, and what is unusable for lumber goes into the chipper.

I think the biggest advantage of the quad is that you can go down to one inch between saws. When edging small logs, you can take a 4x4, a 2-inch, and then a 1-inch. Most conventional edgers can only go down to three inches. Then those extra two pieces must go back to the resaw. With the quad, it is done in one pass.

Part four by Richard L. DeArmond

We have been talking about the upstairs flow of the sawmill, or what might be called the board flow. But the Rube Goldberg idea that I would like to discuss concerns the downstairs flow, where the debris and waste go out the other end.

Like many other companies, we ran into what I call a neighbor problem. Here is how it happens: someone builds a sawmill on the back forty. Then the crew thinks, "Well, it would be nice to live right close." So one man builds a house and pretty soon another one. Soon there are 30 or 40 houses. Then their in-laws come along and start moving in. So someone else says, "Well, I ought to build a store here. There's lots of houses." So he builds his store. Pretty soon, this conglomeration becomes a community, with a school and a hospital. And the sawmill is right in the middle.

Then the community gets together and has a meeting—and, by gosh, the mill is a nuisance! That is just the situation we found ourselves in.

In searching for a solution, we spent a lot of money on our burners. We put on controls and blowers, but none of them quite did the job. At the ecology meetings—also called waste disposal meetings—we ran into a man who had a barrel down in southern Idaho. He asked us to send him a couple sacks of our refuse. He said, "I'll throw it into my barrel and it will lighten the sand and dispose of it."

That sounds pretty Rube Goldberg to me. So we did this. We checked the idea out. The man—Mr. Norman Sowards was his name—was a well-educated man in the combustion field. He showed us the basics of his idea and we decided to try it with a test model.

The process starts with a vessel, which is a tube 14 feet in diameter by about 22 or 24 feet in height. There is a dead air space—the lower plenum. There is a sand base, which is the burning bed. There are no grates.

We had quite a time preventing the sand from coming through the brick plate. But in the system now, the bottom of the bed is dead and just the top of it moves. This keeps the sand from sifting down through the holes.

Our chief problems have been with fuel injection. We finally used forced air to disperse the fuel. Temperature in this area is 1800 to 2000 degrees. As it falls in, most of the fuel will flash out. But the big chunks fall in the sand bed. The air coming in at the

Hogged waste at Idaho Forest Industries' Coeur d'Alene, Idaho, plant moves downline toward the fluidized-bed burner seen in background.

bottom fluidizes the bed, so it is just like quicksand. Temperature of the bed varies from 900 to 1800 degrees. As the material goes in, it disintegrates.

The first model just had a stack at the top. The gas coming out at 2000 to 2200 degrees had so much velocity that it was carrying some particles out with it. So we decided to go through a common cyclone. That cyclone takes all the particles out and dumps them in an ash bin. Then the hot gas goes out the top.

This system is just like a burner disposal system, so we have to cool the exit gas. We try to keep it below about 800 degrees. Otherwise, it would burn out the cyclones. The largest particle that goes out the top is 30 microns.

At the Atlas mill, we were forced to install a system. We had decided that, if we were ever going to make it work, we would have to shut the burner down. So we put the fire out and immediately filled the burner with refuse. Then we did not dare light it. It has been full for two months, but we have been disposing of all the refuse from the Atlas mill and part of the refuse from the stud mill by the new method. But the fuel must be

A closer view of the fluidized-bed burner unit, which burns at 1,800 to 2,000 degrees Fahrenheit, causing flash destruction of most refuse. The unit has won strong encouragement from Idaho air quality authorities.

The ash bin collectors beneath the cyclones, which were installed to capture any remaining particles ejected by the updraft resulting from the 2,000- to 2,200-degree stack temperature. Perhaps three-quarters to one yard of ash is produced per week with the burner in 24-hour, 7-day operation.

prestored and it needs to be metered out over a 24-hour period. So we have a 60-unit storage bin ahead of the vessel and meter the fuel in over a full 24 hours.

I suppose people wonder about all that heat going out the tube. We are working on that. York Shipley has designed a boiler package to go on it. Our old burner was a 50-footer with a 20-foot extension. We tried many methods to make it work. But I think we have done the best thing now by filling it up.

DISCUSSION PERIOD

QUESTION: Dick, are you burning bark as well as wood in this burner?

DeARMOND: We are burning bark, sawdust, and shavings. We have even tried garbage in it. It requires a hog in your line, because the fuel must be hogged.

QUESTION: Many of us have problems burning bark. Much of that problem centers around slag—sand, grit, rocks and boulders, deer horns and guts, beer cans, and everything else. What kind of a slagging problem do you have in a fluidized burner?

DeARMOND: We really do not have a slagging problem with the burner. What little there is goes out as ash and is collected in the bin at the base of the cyclone.

QUESTION: Larry, how often do you change saws on your quad?

WEBB: We usually run the two saws next to the fence for four hours. The two saws on the outside of the fence we usually run for about eight hours. Of course, we have the best saw filer in the country, so that makes it possible for us to run for four and eight hours.

QUESTION: Dick, what type of fuel do you use to start the burner up, and how much do you use a month?

DeARMOND: It is necessary to preheat the bed to round 900 degrees, although wood will combust at about 750 degrees. We are using propane in it now, but are about to hook into the natural gas line. After the burner is in operation, it is self-sustaining unless the bed temperature is lost.

QUESTION: Dick, how much ash do you accumulate in a week?

DeARMOND: I cannot give you a specific amount. It is about three or four bucketloads with the tractor's ¼-yard bucket.

QUESTION: Have you run any flow rates, Dick? Do you know the maximum unit-per-hour rate you can sustain with the burner? Right now is this scaled to your mill or is it still in the prototype stage, with future expansion planned?

DeARMOND: Right now it is still in the prototype stage. We are running the prototype at the Atlas mill, and we can run upward of 10 units through it an hour.

QUESTION: This is a 200-cubic-foot unit measure? At 10 units per hour?

DeARMOND: That is correct. A moisture content of 60 percent is where you begin to run into trouble. Then you have to put some shavings or drier fuel in with it.

QUESTION: After your experience, Dick, how do you feel about planning and zoning?

DeARMOND: We are right in the middle of it. If you are serious, I think it is the only alternative.

14

Careful lumber handling cuts degrade losses

Dean W. Huber, Huber Consulting, Lake Oswego, Oregon

Various types of applied technology—with the objective of increasing recovery of available fiber—are available for today's sawmills. This applied technology is in log scanning, computerized sawing, data collection, and process evaluation. But now that we know everything about the raw material and how to convert it from a round log into a rectangular board, we must not forget about the boards. Logs and labor cost a good deal of money. *The important thing is to get logs through the mill and lumber into the boxcar with maximum recovery.* If we cannot keep track of our lumber and then sell it, there is no point in producing it.

Mother Nature provided a wide variety of degrade factors in normal wood. Wane, rot, knot, spiral grain, ring shake, sinker stock, wet pockets, and so forth, provide enough problems to keep us busy. Every profit-oriented sawmill manager is looking for better ways to maintain lumber quality and volume. Man-made degrade is not needed. So let us look at some common, everyday methods of maintaining lumber quality by plugging the dikes on degrade.

Over my years of travel, I have taken many pictures. These are put together into a lecture of two parallel programs which compare and contrast poor and good milling practices. The photograph at the left or top of each pair shows poor mill practices. At the right or bottom of each pair is a contrasting picture showing good mill practices. This chapter will examine several methods of reducing common causes of degrade through improved sawing, trimming, sorting, stacking, and drying methods.

A word on thickness

Several Clinic speakers have talked about precision-this, precision-that. They talk in terms of thousandths of an inch, overrun on log scale, increased LRF, and so on. In addition, you should follow the disadvantages of thick lumber through drying and planing. You will find that thick lumber is extremely costly. Thick lumber dries more slowly than thin; it does not keep the stickers in place, and it blocks the air flow through the packages. Thin lumber, as well as on-size lumber, is then dried too long and is allowed to warp—producing degrade.

If we have uniform size cutting—both in thickness and length—we can pick up a tremendous number of benefits in lumber quality, in drying costs, and in planer production.

Uniform trimming prevents waste

Many high-production mills, particularly stud and dimension mills, have real problems with trimming. The piece count is too great for the trimmer to handle adequately. If you are concerned only with volume and/or piece count, you soon produce a problem with grade (Figure 14.1, left). From the kiln operator's standpoint, that extra 2, 6, or 8 inches is very undesirable in terms of gaps created between the loads. Proper end-trimming permits humping kiln car loads together very closely (Figure 14.1, right). By closing these gaps, air circulation is forced through the loads of lumber by preventing short-circuiting of air around the ends. Poor end-trimming, raising the trimmer saws, and/or bypassing the trimmer are only short-term solutions. Mill production is maintained, but problems begin to compound downstream at the dry kilns, planer, and dry-end trim saws.

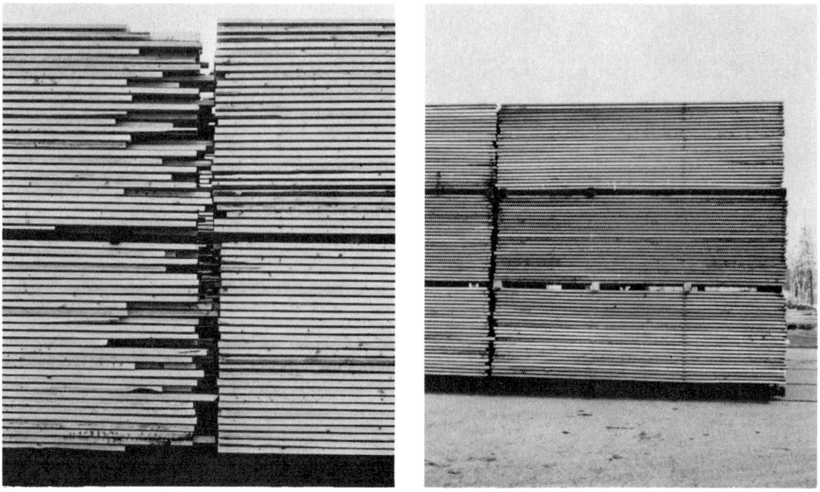

Figure 14.1. *Left*: Imprecise end-trimming causes gaps between kiln loads, letting air short-circuit around the ends of the loads. *Right*: Accurate end-trimming permits bunking loads closely together, improving air circulation and reducing kiln-generated degrade.

Sorting to length

We could devote an entire Clinic to sorting—for grade, thickness, species, and so on. In this chapter, I will limit my comments to sorting to length, or to length group. Two common problems are letting the wrong pieces into whatever length groups are established and setting up incorrect length groupings.

At one mill, sorting was over a device designed for 70,000 board feet per shift; however, the mill was putting 150,000 board feet over it. You can see the consequence in Figure 14.2, top. The sorter was not capable of performing at the required production speed. We improved the machine and attained 20 percent more lumber per kiln charge while reducing drying time, drying costs, and trim loss. When carefully sorting to length, you can get the type of quality stacking required at the dry kiln (Figure 14.2, bottom).

Sorting to length group can become a problem, however. For example, in a given mixed-length group—18s and 20s, for instance—there may be too few 20s. Very few of the long-length boards will be recovered as 20-footers on the dry chain. They will

Figure 14.2. *Top*: When sorting to length, be sure the equipment can handle the volume required. Otherwise, odd-size pieces will occur in stacked loads, negating the value of sorting to length. *Bottom*: Careful sorting to length allows high-quality stacking, increased production, and improved quality.

probably degrade due to end-split, cup, or twist. It may be necessary to trim them to maintain grade. So the extra length is lost anyway! You might just as well have tried to find a method of sorting to length and, in addition, gained extra dry kiln production and capacity. I know that certain length sorting is necessary—for example, with 14s and 16s. My pet peeve is sacrificing two feet of kiln space per car to accommodate only four or five long pieces. The volume of a two-foot void per car, in terms of the entire kiln charge, becomes very significant. When there are only a few long lengths in a group, find someplace to sort them out and prevent their degrade.

The benefits of proper sorting will be observed downstream at the kilns, as illustrated in Figure 14.3. The top photograph shows poor sorting to length with long ends hanging out. These ends are exposed to extra air flow and dry very quickly; they also dry very crookedly. These boards have twist, end-split, and all sorts of degrade that the grader does not like. Since the grader is working for you, he will cut the ends off to upgrade the boards. This will lose the volume and overrun produced in the mill. By sorting to length, as in Figure 14.3, bottom, air flows are controlled and stickers can be placed at the ends of the lumber where they belong. This will restrain the warping of boards by weighing them down, board upon board. The result is to minimize end-check, twist, warp, and other problems, while improving lumber quality.

Proper piling for optimum air flow

The objective of piling is simply to separate the lumber, allowing it to dry. Circulating air in a dry kiln is your workhorse. Do not mistreat it. If you have the kind of condition that pinches off air flow (Figure 14.4, top), air is not going to flow evenly through the package. Allow that air to pass smoothly and uniformly through the courses of lumber (Figure 14.4, bottom) and it will do its work for you.

Stickers should be aligned vertically. Stickers not only separate the lumber, they must also distribute the weight of lumber from top to bottom—through the spacers and bunks—down to the kiln cars. Improperly placed stickers can be very costly to lumber grade by causing kink, twist, warp, and ski-tipped ends. The consequence of improper placement is downgrade of board value. Keep in mind that every sticker which is out of line affects two courses

Figure 14.3. *Top*: Ends of long boards stacked with a shorter length group take up unnecessary kiln space and incur degrade factors during drying. The additional length will generally have to be trimmed to upgrade the board. *Bottom*: Sorting to length permits the stickers to be placed on the ends of the load, where they can minimize degrade during drying.

Figure 14.4. *Top*: Improper placement and failure to vertically align stickers affect two courses of lumber—above and below each sticker. *Bottom*: Careful sticker alignment provides necessary support (preventing kink) and permits good air circulation through the lumber.

of lumber, i.e., the course above and the course below.

In package operations, sticker alignment assists the swamper and forklift operator in positioning the 4x4 spacers between packages. If the stickers are in line, spacers can be accurately positioned. If the stickers are not well aligned, and the spacers thrown in at random, degrade will be created in the entire top course of lumber. If the packages are piled 20 courses high, then 5 percent of the lumber in that package may be afflicted by misplaced spacers. What is going to happen? Degrade is caused due to kinks, bends, twist, end-split, cup, and so on. Your grader and trimmerman will try to upgrade that stock, and the consequence is a loss in wood volume. The value of a 2-foot piece in the waste conveyor cannot compare to the value of that piece if it remained on a high-quality board. The lesser of two evils is chosen when you have to end-trim to save grade. Chip values are less than 10 percent of the board value.

Forklift stacking requires skill

Stacking operations are generally referred to as those done with the forklift. The forklift driver is more important than many people realize. He can do a lot for you or a lot against you. Unfortunately, foremen often simply say to him, "Your job is to pile the loads up vertically," and go away. If he is a good driver, he may figure out what is required. But if he is not a good driver, and he is not thinking, you may end up with the kind of condition shown in Figure 14.5, top. As you examine this picture, observe that the forklift driver has not had help in the stickering of the lumber as a package.

We can begin to see a consistent pattern. When the mess upstream is continually passed downstream, the attitude of the people follows right along with it. So plugging these dikes upstream simultaneously takes care of many problems in the rest of the operation. In stacking, we want good vertical alignment as in Figure 14.5, bottom. This affects air flow in the dry kiln, weight distribution from board to sticker and package to package, drying rates, and gas consumption costs. In addition, the lumber going out to the planer is a better quality product.

Sometimes the forklift driver is not given a chance. He gets a

Figure 14.5. *Top*: Forklift operator failed to align lumber packages vertically. But the lumber packages themselves show the results of careless operations upstream. *Bottom*: Kiln loads show good vertical alignment. Forklift operator was assisted by careful preparation of packages upstream.

mess and it is his job to put it someplace. In the process, he drops a few pieces off the inside, which block the air flow from package to package. The result is uneven drying with poor lumber quality, and no one can figure out why. Sometimes a forklift operator drops a board onto the concrete or into the mud and then rolls over it with the lift truck and it ends up as pulp, not high-grade lumber.

Another place where the forklift driver is important is his ability in piling packages next to one another. Keep in mind that the only thing working for you in the dry kiln is the air that passes through the courses of lumber. If that air cannot get from one package to the next, it will not do any work. If the air is blocked by an abutting package, a dam is set up and the lumber cooks like a turkey in an oven. Given enough cooking time, it will dry, but only at the cost of overdrying properly positioned packages. Drying costs go up, and grade value goes down. This is why, in a forklift operation, we like to leave a couple of inches of chimney—which, in essence, becomes a manifold—between the two packages. Air passes through one package of lumber, then regroups itself before passing through the second package. The chimney permits air to move between packages without interruption—particularly where stacks are not of uniform height.

I once worked with a plant manager who complained about wet lumber repeatedly, year after year. He called in the kiln sales representative, the kiln engineer, the local Forest Service personnel, and even a professor from the local college, but to no avail. The problem was too obvious to be evident. The forklift driver thought he was doing a good deed by pushing the packages together to keep them from falling apart in the kiln. And so he was—but he was also causing a wet spot problem by blocking the air flow. The solution was simple and cost no money: retrain the forklift driver to allow a few inches between packages.

Another crucial area that the forklift driver should watch is vertical alignment of sticker tiers over the kiln car bunks and placement of spacers between the packages. These two areas probably produce the most consistent and most voluminous degrade losses in our sawmill industry. Kink created at these locations is uncannily consistent. Lumber with severe kink is rejected ahead of the planer and becomes a total loss. In less severe cases kink may downgrade the lumber, even if it remains within the grading standards. Consider the contractor building a house. He prefers to work with flat lumber rather than kinky lumber because his costs of materials and labor are less.

Flat lumber is rather easily attained if we maintain sawing sizes and vertical alignment of packages, stickers and spacers. The advantages of proper practices are illustrated in Figure 14.6. These are "before" (left) and "after" (right) pictures of the same mill. The right-hand photograph shows seven sticker tiers, seven spacers, and seven bunks across the bottom to support the weight. Notice that the kiln cars are stacked four packages high instead of the usual three high. In order to distribute this additional weight, we put in the extra stickers and spacers. It worked! Lumber quality was maintained.

Kiln maintenance

Let us turn to drying. The rough sawn lumber is put into a box (the dry kiln) where we perform some voodoo—hoping that it comes out flat and evenly dried. If the lumber coming out of the kiln is not what we expected, the drying schedule could be at fault. As mill managers, we may not be top-notch kiln operators,

Figure 14.6. *Left*: Misalignment of sticker tiers and improper placement of bolsters between packages are perhaps the two most prevalent causes of degrade. *Right*: Sticker tiers are vertically aligned with the spacers and kiln car bunks. Extra stickers support weight of the fourth lumber package.

but we should inspect the kiln on occasion. If we see a chart with irregular lines similar to that in Figure 14.7, top, it ought to tell us that the kiln is not in control. Bring in the kiln operator and ask him the cause; he may already know. If not, bring in somebody who is capable of interpreting the mess and getting it rectified. On the other hand, if we see a chart similar to the one in Figure 14.7, bottom, it tells us that the kiln is basically in control. You may not understand the schedule, but at least it is comforting to see that the pen lines are straight.

Just as important is kiln maintenance. Assuming that you have a capable kiln operator, he cannot perform properly unless he has the right kind of equipment. Note the condition of the kiln in Figure 14.8, left. Several ceiling baffles are gone, floor baffles are damaged, and the end baffles are rusted into position from lack of use. There is water running in at the bottom of the kiln. Much heat is being lost because of the poor quality of kiln coating. Preventive kiln maintenance is something every mill manager should adhere to–lack of attention can result in significant dollar losses.

The dry kiln in Figure 14.8, right, has baffles in position. This shows that if the ceiling baffles are down and the end baffles and floor baffles are in position, then the circulating air has only one place to go–through the lumber packages. Baffles prevent air from escaping over the top, around the end, or underneath the load. I recently checked a kiln in which only 25 percent of the air was employed for drying; the other 75 percent was bypassing the loads. What a waste of energy!

Savings estimated

So far we have considered some ideas for plugging the dikes on degrade. Now let's talk about how much money, production, or recovery you can expect to gain with these techniques. This data was collected from our past experience. Your actual numbers, of course, will depend on the quality of your current operation.

In some mills we have generated over 35 percent more volume through the dry kilns. This was accomplished by: (1) minimizing or reducing problems caused by missorted lumber; (2) getting more lumber into each kiln charge; (3) developing shorter drying

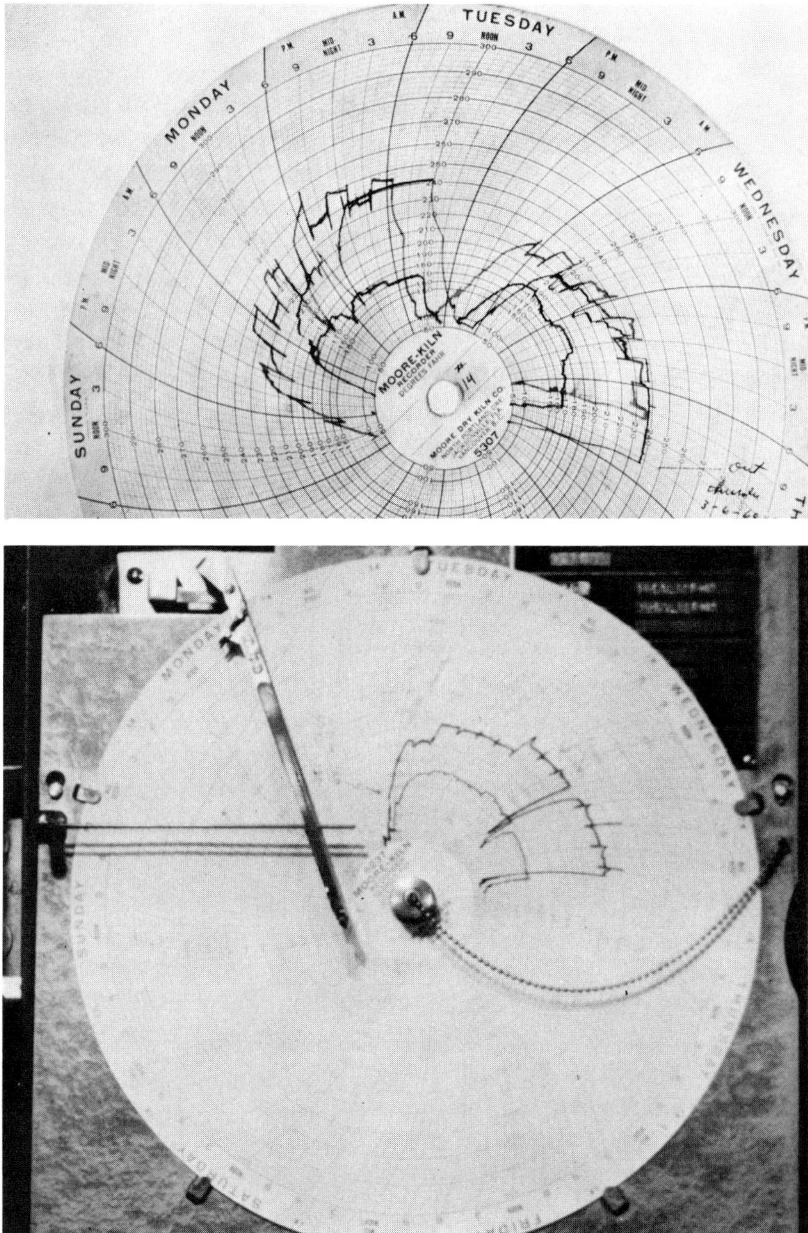

Figure 14.7. *Top*: Jagged, erratic lines on the dry kiln temperature chart show an out-of-control situation. *Bottom*: Temperature chart readings should show this type of consistency.

Figure 14.8. *Left*: Much heat is being wasted in this kiln because of poor baffle conditions, ill-maintained kiln coatings, and water running into the kiln. *Right*: With end and floor baffles in correct position and ceiling baffles down, air is forced to circulate through the lumber stacks.

schedules. The result was a 15 to 20 percent reduction in trim loss and 40 percent lower drying costs.

By reducing plug-ups at the dry end trimmer and the number of pulls on the dry chain, planer production will go up and operating costs down. One mill reported a reduction in planer costs of about $1 per thousand board feet—attributing it to a conversion from random length sorting to group length sorting.

Improvements in piling can be realized by placing stickers where they belong. At one stud mill this yielded a 7 percent increase in stud grade recovery and a 10 percent increase in planer production. Improved stacking yielded about 5 to 10 percent increase in grade by minimizing kinks caused by the spacers and bolsters.

At the dry kiln there are many maintenance items which can yield savings. For one mill we simply corrected a minor instrument defect and cut drying time by 10 percent. If there is any single investment worth making in a dry kiln, it is in baffles. These will keep air going only where you want it to go—through the lumber. In one mill, simply installing floor baffles reduced drying time by about 10 percent.

A great deal of research needs to be completed on drying schedules. As a wood technologist, I can assure you that there is much room for improvement in this area. But, until this is done, there are many areas throughout the sawmill where the techniques discussed here will help minimize dollar losses caused by unnecessary lumber degrade. This is what I mean by—plugging the dikes on degrade.

15

Noise control: the compliance options

Frank N. Berglund, Safety Engineer, Roseburg Lumber Co., Roseburg, Oregon

This chapter will cover five points on noise control compliance:

1. Brief review of the history of the law from the Walsh-Healy Act to the present-day formulation of the Occupational Safety and Health Act (OSHA) standard.
2. A view of the three alternatives for complying with the law: engineering control, administrative control, and hearing conservation program.
3. The practical, feasible approach to the alternatives.
4. Brief outline of the final draft of the standard as proposed by the advisory committee to the Department of Labor.
5. Noise control activities of the trade associations, including a practical solution to joining forces in noise engineering control. (OSHA is reportedly quite pleased with the work the industry associations are doing in this area.)

Enforcement of the noise control provisions of the Walsh-Healy Act led a number of the forest products companies to meet in

Portland, Oregon. After seeing what had happened with another Northwest company, we at Roseburg Lumber Co. felt we were next in line. So, we helped organize a meeting to discuss the situation and decide what to do. At that time we called our group the Northwest Noise Committee. The consensus was it would be difficult to obey the law but that we really had no choice, since it was clear the government would be tough in enforcing it.

At the second meeting of the group, about 60 companies were represented. A number of them volunteered to work on noise control engineering techniques. Companies selected particular pieces of equipment in the sawmill, the plywood plant, and other operations. When we held our first seminar and presented the information developed to the public, it was very successful. In the second year we got further information and put together a high-quality package of noise control information, in the form of a slide presentation, available to the forest products industry through trade associations.

The young government inspector who visited our plant had been with the Oregon State Board of Health previously. He knew our plants and where he wanted to go to get things rolling. And he made his noise inspection.

We felt that we should not get too worried until we had had a chance to take a good look at the situation. One thing we did was to find an instrument that measures noise exposure on a time basis at the different levels specified in the Act.

After taking some measurements in our plant, we found some places that were clearly in violation. We found others, however, where the instruments showed we had less than the maximum allowable hours of exposure at any of the levels in the timetable.

In our analysis and report to Walsh-Healy, we told them our findings. In reply, they said they would accept our findings with our instruments for the time being. In the meantime we found out we were not affected by the Walsh-Healy Act since we had no government contracts.

But this was a preview of what was to come—the Occupational Safety and Health Act.

Members of the Hearing Conservation and Noise Abatement Committee of the Wood Products Section of the National Safety Council, which I will refer to as the Racket Squad, put together a program, which we were invited to present before a Southern Forest Products Association safety conference in New Orleans in February 1972. The program presented in New Orleans incorporated industry input to the slide and tape presentation referred to previously. The southern wood products industries, not having been active in the noise control area and aware that they would be involved eventually, responded with great interest to the slide program made available through their association.

Alternative methods of compliance

With regard to legal liability, we hope to develop a means whereby an operator can stand up in court and say, "This man could not have suffered his hearing loss in our plant. We can prove by records that we are well below the levels which supposedly cause hearing loss." This is ideal. But how to do it?

There are three different ways, one of which is engineering control. With the right materials, the machine can be housed so that neither the operator nor anybody in the area will be affected. This is the ideal solution. Another idea is partial isolation of the machine—or of the operators in the area from the noise—by using bulkheads. It is also possible to house the operator. In fact, it is already being done on some jobs. As we develop these methods throughout the industry, however, it no longer will be possible to say, "it is not feasible," and forget about it. We will be expected to solve our noise problems.

But there are also many areas where partial or total isolation is not feasible. An example is our new planing mill. There are three new planers; two are in now and another is being installed. It was a beautiful job of housing. The housing lowered the noise level of the planers to an acceptable and nonharmful exposure for those working in the area. But we come back to the problem of feasibility. Feasibility, in this instance, refers to the fact that, although we spent the time and money to reduce the planer noise level, we then found that machine noises in the 95- to 96-decibel range produced by equipment other than the planers took over. Engineering control is not feasible for these noise problems; so, although we controlled the planer noise, our employees still have to wear hearing protection.

The second method of compliance is administrative control. A man in the 96-decibel area can take only four hours at that post a day. Then he must be moved to another job well under the noise level limitations for hearing loss. This is administrative control—moving people around. Is this feasible? I am not certain. If it is being done, I would like to look at the operation. I cannot see our company doing it. Union contracts and wage rates for different jobs make it practically impossible for us.

The third alternative, hearing conservation, is here to stay. Hearing conservation means wearing proper hearing protection. What is proper hearing protection? Many different plugs and types of ear muffs are available. We stock several types. Mill managers know the reaction in the mill. It's "I can't wear this one" and "I can't wear that one." Despite the protests, find something the men can wear and insist they wear it in noisy areas.

Oregon had quite a good law and standard before Walsh-Healy or OSHA had anything on hearing conservation, such as audiograms. The Racket Squad met in Portland and, through trade associations, we assembled our ideas about what we thought was wrong with the OSHA standard and what some of the dangers

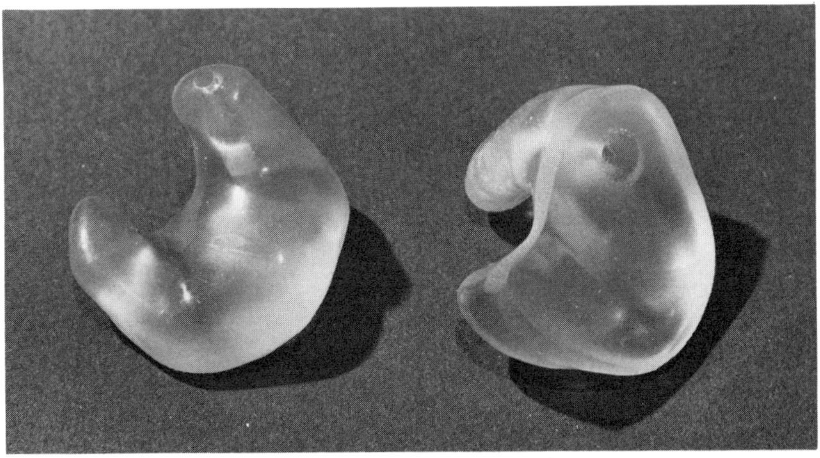

"The Noise Braker," a custom-fitted ear "defender," limits all entering sound to 80 decibels. Tiny valve, which shuts down when noise level rises, permits wearer to hear normal conversations. It is molded to the individual ear, eliminating the discomfort of a rigid plug. Manufacturer is Hocks Laboratories, Portland, Oregon.

were. We presented it to Ken Patrick, safety director for the Western Wood Products Association. Ken presented our ideas to the advisory committee to the Department of Labor. And we did get many of the things we asked for.

I do not know how many of you have dosimeters or exposure meters. We have studied some extensive and expensive sound measuring devices. Based on my professional knowledge and handling of Workman's Compensation claims, however, I caution against depending on one of them as a defense in court.

For instance, several years ago at a meeting addressed by a very prominent Portland attorney, an insurance executive asked him, "What are you going to do about our baseline and our audiograms when we have our records and we have a good conservation program and we go to court here?" The attorney replied, "We'll do everything we can to discredit you." So it will take more than an exposure meter to show compliance with the law. With the Workman's Compensation situation, the primary legal responsibility and burden of proof rests with the employer. Any reasonable doubt as to whether the act of the employee, at the time of the injury, is contemplated by the employment, must be resolved in favor of the employee.

Is there a practical and feasible approach to the three alternatives? Some approach exists to all three, with the possible exception being, for our company in particular, administrative control. But in other industries (and possibly in some forest products industries) administrative control is feasible.

Advisory committee proposal

The proposal that will be presented to the Department of Labor from the advisory committee says:

> Specific provisions of this standard shall be revised as necessary in view of additional information which may become available. Moreover, the standard shall be given a complete review at least every five years and changes made commensurate with increasing scientific knowledge relating to the effects of occupational noise exposures and aspects of noise measurement and control.

This is good for us. A study is underway that, we hope, will help forestall lowering the noise level to 85 decibels in the near future. It may delay this at least five years. We need a chance to do more study. It may be lowered to 85, and even 80, some day. So, when doing engineering control today, it is better to work for 85-decibel levels. If you succeed in lowering the level to 88 decibels, or a gray area, and it is accepted for the present, consider what it will cost to revise again later.

Trade associations active

How active are the associations in this? One woodworking association hired a firm to study various plants and to arrive at a package to help them. We in the Western Wood Products Association plan to engage expertise from a well-known U.S. firm to examine sample member plants. This study should produce a cross section of the real problems in all plants and with the different machines in operation. We want some answers as to what we can do material-wise and what we can do with this or that type of machine. We also will ask the firm to prepare evidence on feasibility for presentation to legislators.

At Roseburg, we had an OSHA inspection and were cited for 36-inch-high guard rails. We have miles of them in that huge flakeboard plant, around the chip bins and elsewhere. The guard rails met the Oregon standard for many years and we feel it was a good code. We also were cited for some very isolated steep stairways, which our owner affectionately calls "ship's ladders." We asked for a variance. With the help of Ken Patrick, we took photographs and we turned the case over to our attorneys in Portland. Patrick was very instrumental in getting our variance granted. The variance was granted on proof of lack of personnel traffic, as allowed for in the standard. We had hoped that, in light of our variance, a grandfather clause might be added to the standard so that all industries would benefit. No grandfather clause was added, but information from OSHA indicated that other firms might apply for, and be granted, a similar variance. Since our citation and later variance, however, we have been putting in the required 42-inch railings at the Roseburg installation and meeting the OSHA standard.

DISCUSSION PERIOD

QUESTION: What foam insulations are available and are they fire retardant?

BERGLUND: We have used quite a bit of this material. As far as the fire problem is concerned, we are also using vinyl filled with lead with good results. There are many good products that have come along. But I would like to caution that many companies have moved quickly into this field and care is necessary in selecting a manufacturer. Foam and other sound-deadening materials do have fireproof capability. There are many good products today. However, when in doubt about something new, consult others who already are using the product. Also, be sure to evaluate the potential of products you may produce as agents for noise reduction. In our industry we find, for instance, that wood, along with other materials, is practical.

16

Noise reduction techniques for the sawmill industry

Dr. John S. Stewart, Director, Noise Control Services, Inc., Greensboro, North Carolina

Machinery noise in the sawmill industry is a subject of special interest since, under current Occupational Safety and Health Act (OSHA) standards, it is the user who is responsible for violations of noise regulations. Machinery noise will also be an increasingly important consideration in the design and manufacture of new sawmill machinery, since the ultimate responsibility for OSHA compliance will eventually rest with the manufacturers. An increasing number of consumers are requesting noise level information before purchasing new and expensive machinery.

The following discussion presents several proven techniques of noise reduction for sawmill machinery and identifies promising areas of noise reduction in the future.

Sawmill noise reduction techniques

This chapter deals with reduction of machinery noise either at the noise source or in the near field of the source. Thus, techniques such as operator booths, operator barriers, and acoustical treatment of ceilings, walls, and other areas are not considered. These techniques, while often useful in isolated instances, are not noted for substantial noise level reduction. Treated ceilings can reduce the contribution of reflected sound to the total noise level, but they are not effective in reducing noise levels near machine operators.

For the sawmill industry, noise reduction at the source usually involves noise-reducing tooling, while noise reduction in the path requires some type of enclosure. The enclosure techniques—total enclosures, partial enclosures, and baffling techniques—are outlined below.

Enclosure techniques for planers

Several approaches for reducing noise in the path are illustrated in Figure 16.1. The following observations are appropriate.

1. Vibration isolation of the machine itself from the floor is usually effective in reducing low-frequency noise components resulting from floor vibration. It does not usually provide substantial A-scale noise reduction. Note, however, that isolation techniques breaking the path of structural vibration may substantially reduce resonant machine vibration noise.
2. Acoustic baffles are effective in reducing noise by small amounts provided the baffle is located so that the receiver (operator) is in the "acoustic shadow" of the baffle.
3. A properly constructed enclosure should have a sound-reflecting material (mass) in addition to absorptive material to provide noise reduction, since more than one reflection of a sound wave is required for effective absorption.
4. Rigid, sealed enclosures are less effective than enclosures that are acoustically lined and isolated from machine vibration. Isolation may be achieved by the method shown in Figure 16.1 or by isolating the enclosure itself from the vibration source.

Figure 16.2 shows the effectiveness of enclosure alternatives for a noise source radiating acoustic energy over a frequency (pitch)

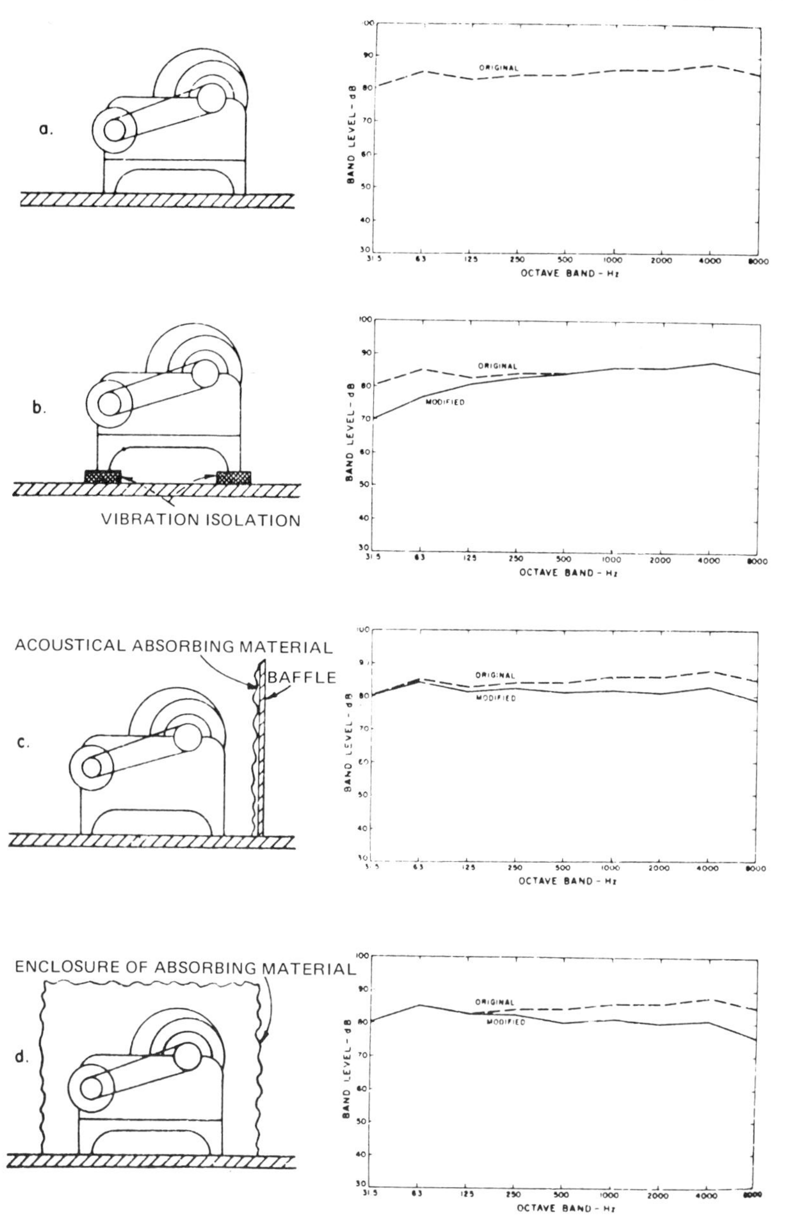

(continued on following page)

Figure 16.1. Noise reduction effects of some noise control measures.

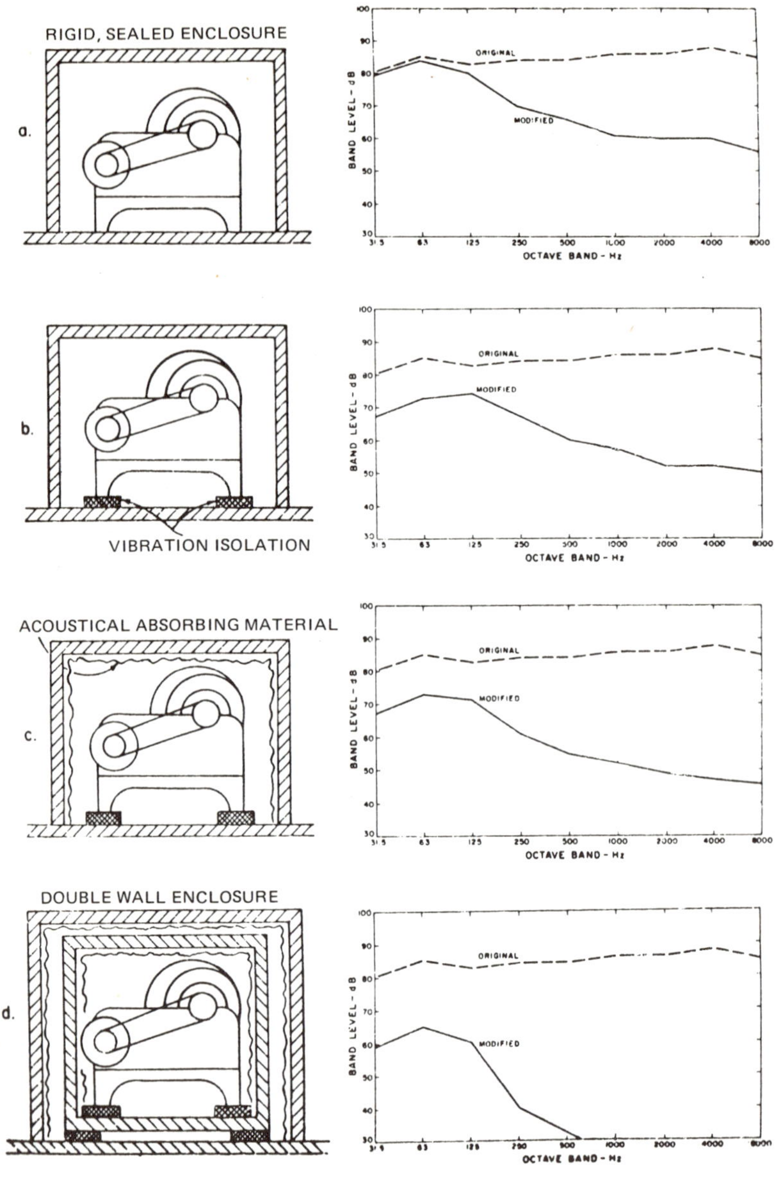

Figure 16.1 (continued)

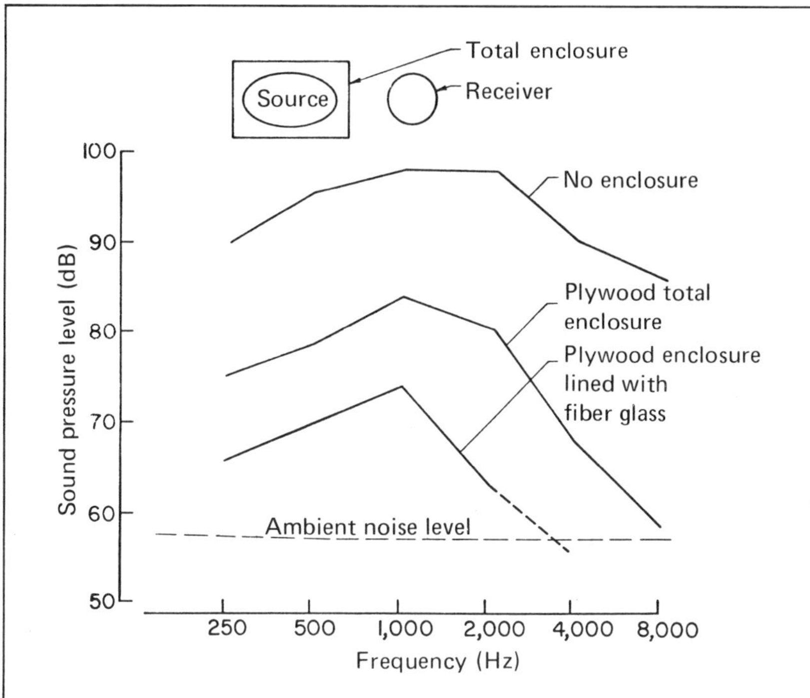

Figure 16.2. Noise reduction using a fiber glass-lined plywood enclosure.

range of 250 to 8,000 hertz. Simple enclosures, constructed of 1/2- to 1-inch-thick plywood lined with 2-inch fiber glass batts are relatively inexpensive and can provide noise reduction of 20 to 30 decibels on the A scale (dBA) provided:

1. The enclosure is properly constructed.
2. The size of both infeed and outfeed openings is minimized.
3. The amount of vibrating workpiece left unenclosed is minimized.

Items 2 and 3 are often overlooked in the design of an enclosure. Tables 16.1 and 16.2 illustrate the importance of these factors. An increase in opening area of 1 to 5 percent of the wall area reduces the effectiveness of the enclosure by as much as 5

Table 16.1. Noise Reduction of Lined Plywood Enclosures as a Function of Opening Area

Opening area (percent of total area)	Fiber glass-treated area (percent)	Noise reduction (dBA)		
		1/2" ply.	3/4" ply.	1" ply.
0.1	25	13.0	18.0	20.0
	50	16.0	20.0	23.0
	75	18.0	23.0	25.0
	100	19.5	24.0	27.0
1	25	10.0	14.0	14.0
	50	13.0	17.0	17.0
	75	15.0	18.5	18.5
	100	17.0	20.0	20.0
5	25	7.0	9.0	9.0
	50	10.0	13.0	13.0
	75	11.5	14.0	14.0
	100	13.0	15.0	15.0
10	25	5.0	5.0	5.0
	50	8.0	8.0	8.0
	75	9.0	9.0	9.0
	100	10.0	10.0	10.0

Table 16.2. Loss in Enclosure Effectiveness Resulting from Workpiece Exposure

Length of board being surfaced (feet)	Loss in effectiveness of enclosure (dBA) board length enclosed				
	1′	2′	4′	8′	16′
1	0	0	0	0	0
2	3	0	0	0	0
4	6	3	0	0	0
8	9	6	3	0	0
16	12	9	6	3	0

dBA. If, in addition, one-half of the vibrating board is left unenclosed, an additional 3 dBA is lost, giving 8 dBA less noise reduction than anticipated. Figure 16.3 illustrates several enclosure alternatives which are currently available.

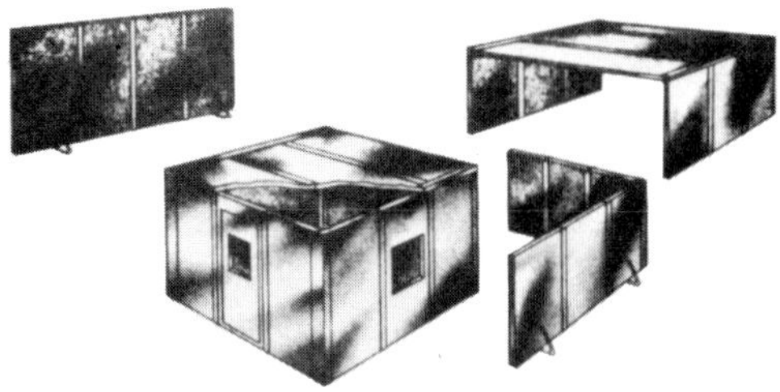

PREFABRICATED ENCLOSURE COMPONENTS

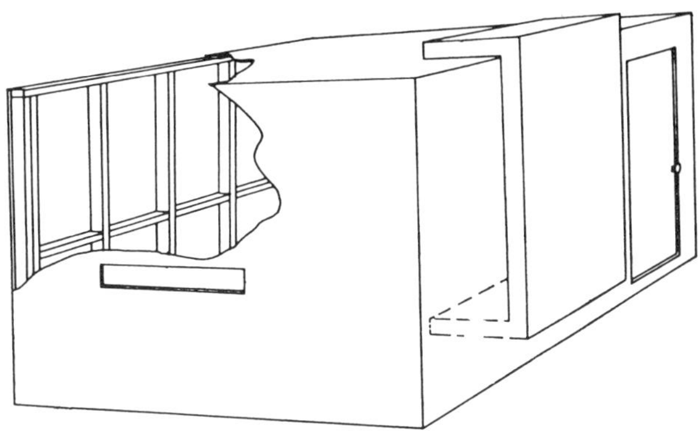

STUD WALL ENCLOSURE CONSTRUCTED AT PLANT SITE

PARTIAL ENCLOSURE FURNISHED BY MANUFACTURER

Figure 16.3. Several available enclosure alternatives.

Noise-reducing tooling

Noise-reducing tooling for the sawmill industry normally includes special cutterheads and saw blades or alterations to the standard tools.

Cutterhead designs. Noise produced by rotating cutterheads results from aerodynamic sources during idle and, usually, workpiece vibration during cutting. Idle noise is created by the periodic passage of cutterhead knives in the vicinity of stationary surfaces near the head (Figure 16.4). Cutting noise results from the dissipation of energy stored in the workpiece as evidenced by the increase in noise level as board width increases, shown in Figure 16.5. This energy in the workpiece results from the periodic impact of cutterhead knives on the workpiece, as illustrated in Figure 16.6. A smoothing out of the force-time history of the workpiece is required to reduce vibration and associated noise.

Factors influencing cutter-knife impact include impact velocity, impact magnitude, and impact nature (Table 16.3). The impact velocity, governed by head revolutions per minute and diameter, is a fixed quantity for a particular machine and operation, and provides a baseline height of the force quantity of Figure 16.6.

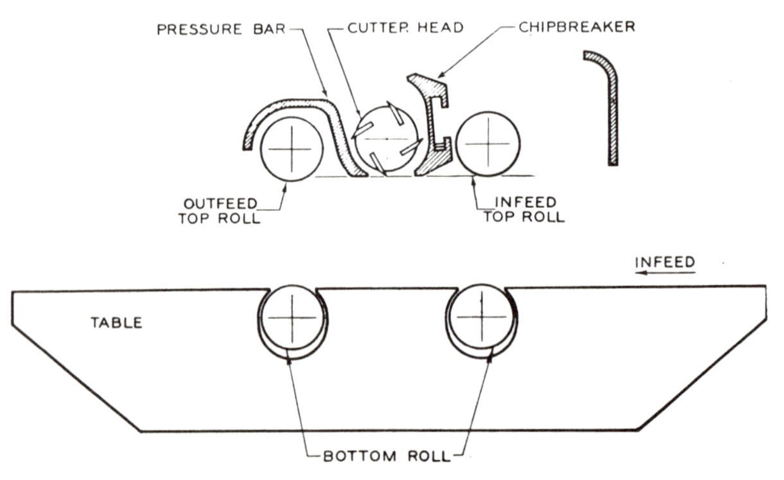

Figure 16.4. Cutterhead location in a typical planer.

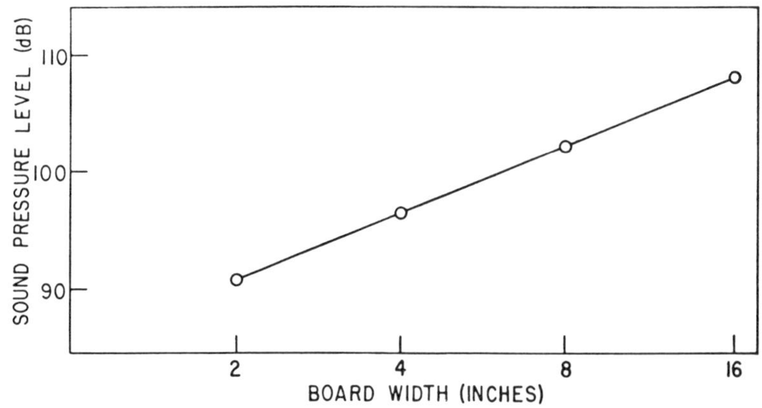

Figure 16.5. Increase in sound pressure level with board width.

The impact magnitude is also affected by variables such as knife sharpness and cutting angles. These factors, governed by maintenance considerations, are also important with respect to noise. The category of impact nature is indicative of the design parameters of the cutterhead itself and reflects any deviations from the standard straight-knife cutterhead, such as helix angle, segmented knife effects, and continuity of cutting.

The most important consideration from a workpiece vibration standpoint is the width of stock free to vibrate, as shown in Figure 16.7. Substantial noise reductions are possible as a result of the exertion of a constant force on the workpiece, rather than the periodic impact of standard cutterheads. The relationship of the width of workpiece free to vibrate (L), helix angle (ϕ), number of knives (N), and head diameter (D) is also given in Figure 16.7. A prototype head incorporating the design criteria of Figure 16.7 has reduced noise by 20 dBA as compared to conventional straight knife heads.

Commercially available segmented cutterheads (Figure 16.8), reduce idle and cutting noise by interrupting the blade passage frequency of straight knife heads, and, to some extent, reduce board vibration by reducing the width of stock free to vibrate.

Noise levels for a high-speed planer matcher equipped with

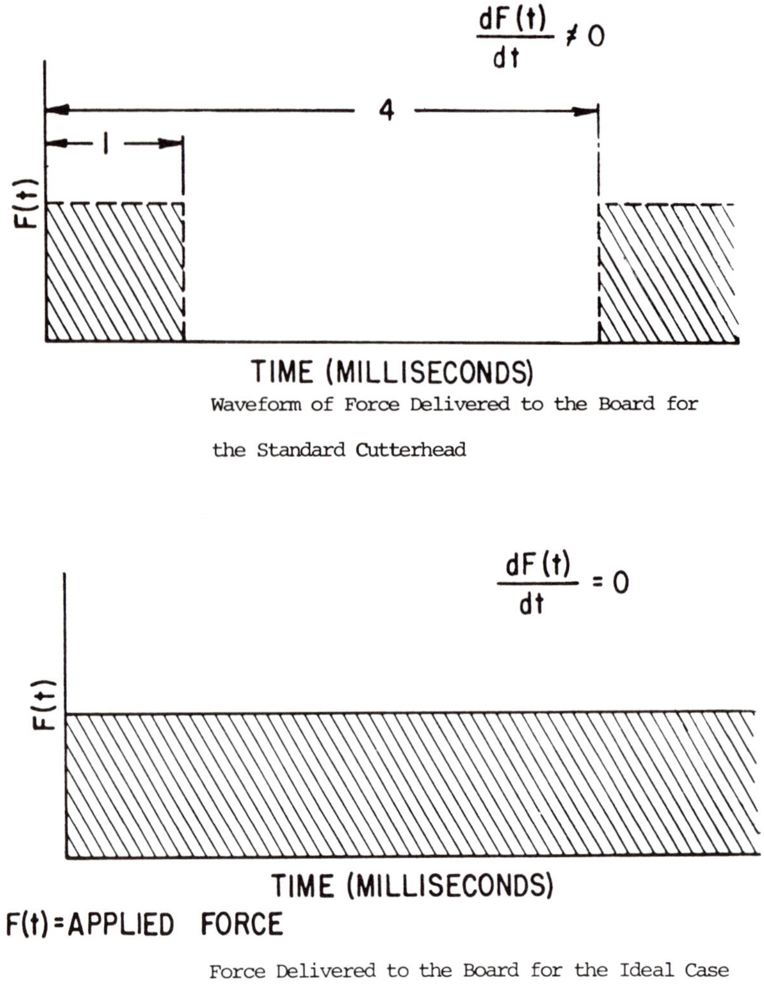

Figure 16.6. Force-time history of workpiece.

segmented heads and a compact acoustical enclosure are compared with the standard machine in Table 16.4. It is observed from the table that a reduction of 8 to 10 dBA during cutting is possible with either a partial enclosure or segmented cutterhead. The

Table 16.3. Factors Influencing Knife Impact

1. Impact Velocity
   a. Revolutions per minute
   b. Diameter
2. Impact Magnitude
   a. Knife sharpness (material)
   b. Gullet shape
   c. Knife thickness and support
   d. Cutting angles
3. Impact Nature
   a. Number of knives
   b. Angle of knives with respect to head body
   c. Continuity of cutting

resulting noise level reduction after installation of both segmented cutterheads and a partial enclosure would have been greater than the 15 dBA reported had the machine drive train noise been less.

Noise reducing cutterheads of the segmented type represent an improvement but cannot, alone, reduce cutting noise to within acceptable limits. It is possible to combine a compact partial enclosure with such a cutterhead to achieve acceptable noise levels, rather than constructing a total planer enclosure.

Circular saws. Circular saw noise usually results from resonant blade vibration or aerodynamic (air flow) turbulence, or an interaction of the two sources. These aero-elastic phenomena involving both types of noise sources complicate experimental parameter studies and account for the failure of various theories and noise abatement techniques to be valid for all saws. This chapter proposes several techniques for reducing circular saw noise. Those believed to be practical for the sawmill industry include (1) reduction of saw blade vibration, (2) reduction of air flow noise about the teeth and gullets, (3) acoustical enclosure techniques.

Several techniques for reducing resonant blade vibration are

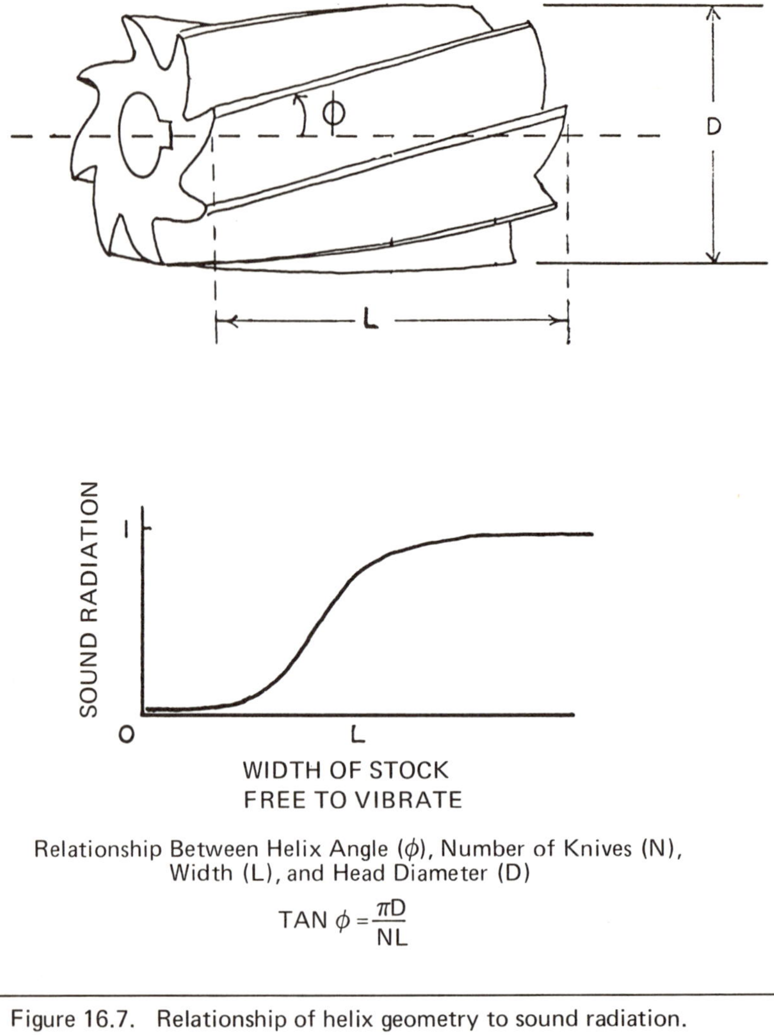

Relationship Between Helix Angle (ϕ), Number of Knives (N), Width (L), and Head Diameter (D)

$$\text{TAN } \phi = \frac{\pi D}{NL}$$

Figure 16.7. Relationship of helix geometry to sound radiation.

given in Table 16.5. The most effective techniques are total blade lamination and the laminated collar approach, provided the collars cover a large percentage of the blade. The practicality of the approaches shown in Table 16.5 depends on the application.

Table 16.6 lists several techniques known to reduce air flow

Figure 16.8. Segmented cutterhead for planer matcher.

noise. The parameters involved and their interaction with each other, however, are not yet well understood. Considerable research will be required before these questions can be resolved. It is observed in Table 16.6 that aerodynamic noise generated by free-running saws can be reduced by reducing tip speed, reducing gullet size, reducing blade thickness, and improving gullet geometry. Tip speed can be reduced by reducing blade diameter or blade rpm, which can also increase saw life, as implied in Figure 16.9 for a nonresonating blade. The slope of the curve (Figure 16.10) is nearly 15 dBA per doubling of tip speed, which means that a reduction from 3,600 to 2,500 rpm could be expected to reduce air flow (idle) noise by 8 dBA. Other techniques listed in

Table 16.4. Comparison of Planer Matcher Noise Levels Using Various Noise-Reducing Techniques

	Standard	Enclosed	Segmented head	Segmented head and enclosure
Idle	103	91	91	91
Cutting	108	99	98	93

Note: Correction to free field = –4 dBA (measured according to WMMA Test Code)

Table 16.5. Techniques for Reducing Blade Vibration		
Technique	**Explanation**	**Effectiveness**
Total blade lamination	Blade consists of two plates bonded together to form the blank.	Blade vibration effectively reduced. Both idle and cutting noise reduced.
Partial blade lamination	Center portion of blade comprised of two pieces. Outer rim is solid.	Blade vibration reduced to some degree. Idle noise reduced due to elimination of blade resonance.
Vibration suppression mechanisms	Physical point contact maintained with blade to suppress resonant response.	Reduction of high frequency screaming noise.
Standard collars	Collars added to increase stiffness of blade.	Screaming (resonant response) reduced in some cases, depending on collar geometry.
Laminated collars	Standard collar with damping layer added.	Effective in reducing noise provided large percentage of blade is within collar.

Table 16.6 can be employed, depending on the requirements of the particular sawing operation.

Enclosure techniques for circular saws

Acoustical enclosures in the immediate vicinity of the saw blade have proven beneficial from both noise and safety standpoints. Table 16.7 illustrates the effectiveness of a partial enclosure used with several different types of saw blades. The steel blade is set into vibration by aerodynamic excitation, as well as direct force input due to cutting. The carbide-tip blades, which are typically thicker than the high-speed steel blades, do not appear to respond as well to aerodynanic forces, but radiate well during cutting. The damped blade does not appear to respond to forces created during

Table 16.6. Techniques for Reducing Air Flow Noise

Technique	Explanation	Effectiveness
Reduced tip speed	Involves reducing blade diameter or rpm, or both.	Impractical in many cases due to required cutting depth and fixed-rpm machines.
Reduced gullet size	Involves selecting blades specifically for the job to be done with emphasis on minimizing gullet size. Usually requires using additional teeth.	Severely limits versatility of saw blade but is reasonable approach in some cases.
Reduced blade thickness and kerf	Involves using thin blades to reduce air disturbance.	Governed by end product requirements.
Alteration of gullet & tooth shape & cutting angles	Initial results promising, but requires considerably more research.	

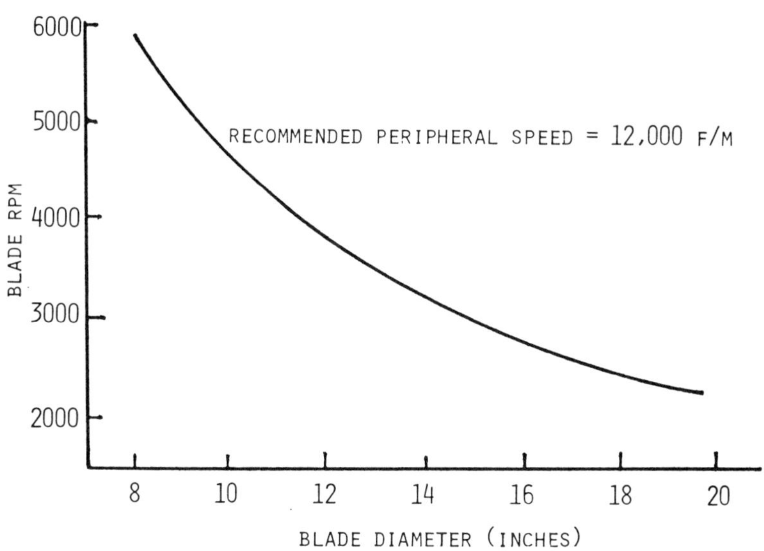

Figure 16.9. Relationship between blade revolutions per minute and diameter for recommended peripheral speed.

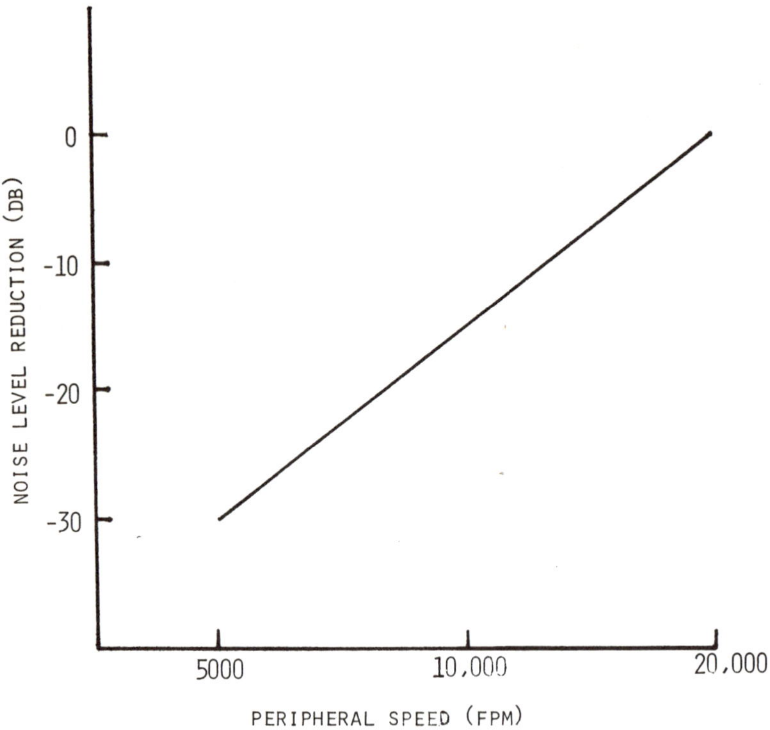

Figure 16.10. Reduction in noise level with peripheral speed.

Table 16.7. Noise Level Comparison of Standard Trim Saw and Saw with Partial Enclosure

	Standard		Lined enclosure	
	idle	cut	idle	cut
	Noise level (dBA)			
Steel blade	105	108	96	102
Carbon-tipped blade	100	105	95	99
Laminated blade	95	96	91	92

idle or cutting conditions. The effect of coating the guard with foam appears to be greatest for the steel blade, since it exhibits the strongest resonant response.

The principle involved in the enclosure techniques presented is near-field absorption of noise generated by the saw. In some cases complete enclosures and simple baffles may prove effective.

DISCUSSION PERIOD

QUESTION: Does balancing the planer head have much effect on noise reduction?

STEWART: Balancing has some effect, but not as much as one might hope. It would not be reasonable to expect a reduction of 10 decibels or more. In my experience, however, an unbalanced planer head may be as much as 4 to 5 decibels noisier than a well-balanced one. Keeping the head balanced and sharp certainly helps–you can pick up anywhere from 4 to 8 decibels. But balancing alone is insufficient.

ROBERT BLOCK, consulting engineer, Seattle, Washington: Has sheet lead been effective in enclosures?

STEWART: In theory, sheet lead is effective. But in my experience, most problems revolve around the size of the workpiece opening. That is the weak link in the system. So the addition of sheet lead on top of plywood and fiber glass or particleboard and fiber glass really is not that much more effective, because the dominant noise source is the hole left open for the workpiece to enter and exit.

QUESTION: What would be the effect of increasing the speed of the planer head?

STEWART: Increasing planer head speed would have a negative effect on noise reduction.

QUESTION: What foam materials are available for noise control and what is their fire risk or rating?

STEWART: The materials commercially available are, to my knowledge, fire retardant. They are available from numerous manufacturers. New companies are entering into the manufacture of acoustical materials very rapidly. Various acoustics magazines list these manufacturers.

QUESTION: How effective are saw collars in noise reduction?

STEWART: Saw collars may or may not be effective. Collars affect the dynamics of a saw blade in various ways. If you have a screaming saw blade, you may be able to drill a hole or plug an expansion slot or do almost anything to upset the very fine dynamic balance known as resonance. Standard saw collars with a damping material added tend to eliminate screaming and often reduce cutting noise, provided a substantial amount of the blade is within the collar.

DUANE C. KENT, Kaibab Industries, Inc., Panguitch, Utah: I first would like to comment on the sheet lead question. I have just finished installing a sheet lead enclosure of about 1,600 square feet around my planer. It effectively reduced noise from about 126 decibels to around 88 decibels. I did not entirely enclose the board going into the planer and the board coming out. I wonder whether I could pick up a little extra noise reduction by riding some soft rubber tires on the boards?

STEWART: Yes, sir. The amount of reduction you would get would depend on the type of feed rolls. If there is enough pressure exerted on the board, then the effect of the board sticking out of the enclosure can be reduced. We have tried a considerable number of suppression systems consisting of a foam-filled tire which was deflected enough so that it produced a flat on the tire as the material was fed in. This is effective in reducing vibration transmitted beyond the tires. The key to the situation is to make sure you cover an area and not just make line contact. For the structural wave length that you are talking about, probably 3 to 4 inches of flat contact area should be sufficient.

QUESTION: I do not know whether such devices are available, but what about the possibility of rubber-coated rolls in the planer?

STEWART: I know of one planer that uses rubber tires as a feed device and this does help. But problems still remain in terms of ability to attain that much flat without getting into maintenance problems and various other feeding problems. We are working on a system of that type now. The obvious advantage is that you can combine one of the partial enclosures with a suppression system to keep the vibration from spreading out over the entire length of the workpiece. Thus you can get away with a partial enclosure.

QUESTION: I have not as yet put any acoustical foam or anything on the inside of my planer. I have only gypsum board, which has just been painted, but I have effectively stopped noise from coming outside through that enclosure. If I coat the inside of the enclosure with some kind of product, will that stop only the noise that is flying around inside? Will it have any additional effect on noise outside the enclosure?

STEWART: When you constructed your enclosure, you wound up with a set transmission loss. Let us say the enclosure is good for a 25- or 30-decibel transmission drop. If you can reduce the noise level inside the house, then you are taking 25 or 30 decibels off a lower initial level. So if you do put some absorptive material inside the enclosure, it certainly should help some, unless you have problems with the openings or other areas. Absorptive material should reduce the noise level inside the enclosure—and this is the level from which the noise reduction of the enclosure begins.

17

Narrow lumber planer designed for smaller logs

Gilbert W. Anderson, President, Stetson-Ross Machine Co., Inc., Seattle, Washington

R. Brian Parsons, Senior Project Engineer, Crown Zellerbach Canada Ltd., Vancouver, British Columbia

Part one by Gilbert W. Anderson

This chapter discusses a new concept in planer design and operation. It also discusses an ultra high speed planing mill that runs dimension, studs, shop, and box—all at high speed.

Stetson-Ross has been manufacturing planer-matchers since the turn of the century. These are large machines with top and bottom cylinders 15 inches to 25 inches wide and sideheads 1 inch to 8 inches high; these machines often have top and bottom profilers capable of running a variety of patterns.

Our planers with these cylinders are running all over the world. Yet they rarely touch a board wider than 12 inches. Feedworks which were designed to run timbers up to 8 inches thick never run anything thicker than 4-inch material in today's mill. In addition, only a limited amount of pattern work is done on planer-matchers today. Well over half the machine is not being used effectively.

For years Stetson-Ross has prided itself on making conventional matchers bigger and bigger and capable of operating at higher and higher feed speeds. Stetson has, and continues to manufacture, heavy-duty planer-matchers capable of sustained runs at 1,200 feet per minute with 20-knife cylinders.

With more and more lumber companies operating small-log mills using advanced, specialized machinery, we felt a need existed for a planer-matcher designed specifically for the small-log generation. We wanted to develop a machine that was unique: one that would be highly productive, simple to operate, easy to maintain, thoughtful toward operator safety, quiet, and that would require a lower capital investment than conventional Stetson-Ross machines. Sounds like the impossible dream.

With these goals, we set about designing what we call the narrow lumber planer. This is not a "stud planer" only. It is a flexible planing system that can be tailored to specific mill situations. The machine comes in three model sizes. One model handles up to 4x4 material; the second, up to 4x8; and a third, up to 4x12. These machines are now designed with a common frame—allowing an operator to "grow" his machine from one size to another and back again, depending on lumber market changes, at nominal cost.

These narrow lumber planers are rugged, compact, highly productive machines with feed speeds from 300 to 1,200 feet per minute and cylinders containing 8 to 20 knives. The machine is simple and easy to maintain. One of the unusual features of the machine is that the cylinders and feed rolls are supported and driven from one side of the machine only. The spindles supporting the rolls and cylinders are turned down from 6-inch-diameter shafts and are tapered from 3 1/2 inches to 4 29/64 inches. The cutterheads have a 12 1/2-inch cutting circle and the feed rolls are 16 inches in diameter. That makes them larger than on any modern conventional planer.

The cantilevered cutterheads on tapered spindles are easy to remove for knife sharpening and adjustment. They can be quickly replaced with a spare slip-on head and taken to the grinding room for sharpening. This results in minimum downtime for knife maintenance. The rolls, also supported by tapered spindles, are much easier to remove than on conventional planers.

Variable pneumatic tensioning is provided on the top feed rolls and top and side head chipbreakers. This provides for positive control of the lumber at high feed speeds. It also makes it easy to raise both the chipbreakers and rolls to clean out breakups.

The open sides of the machine adjacent to the rolls and cutter-

Variable pneumatic pressure adjustments for narrow lumber planer's top roll and top chipbreaker. Manual "up-down" controls are also shown.

Photo by Monroe & Akins

heads are protected with a high-impact plastic, which allows the operator to observe the movement of the lumber through the machine with maximum safety. The open side also makes cleaning up breakups much easier. The top and bottom jointers have automatic stone advance so that the operator's hands never need to enter the cutting area while jointing.

The feedworks have been greatly simplified. A conventional high-speed planer feedworks consists of 14 gears and pinions. The feedworks in the narrow lumber planer consists of sprockets and roller chain operating in a sealed oil bath. There are no gears in this machine.

The narrower top and bottom cylinders, by their very size, reduce cutterhead noise. In addition, sound-absorbing material has been incorporated into certain areas of the machine. We are working on new cylinder designs to reduce noise further.

We were fairly well along with the conceptual design of the narrow lumber planer when we started discussing it with management and engineering representatives of Crown Zellerbach-Canada. After several meetings with engineering and operating people, Crown decided to install a machine in its new Armstrong, British Columbia, mill. This machine has been operating since the end of 1973. We believe we have met the goals originally set for this machine, including lower capital investment.

The film shown at the Clinic of the machine at Armstrong shows a Stetson-Ross Model 6-10-A1 planer, which was designed in 1945, running adjacent to the narrow lumber planer.

This gives a good comparison of a conventional 30-year-old planer design with the narrow lumber planer. The 6-10-A1 planer is operating at 450 feet per minute with eight-knife cylinders, while the narrow lumber planer is operating at 800 feet per minute on random-length 2x4s. Because of the 16-inch roll size of the narrow lumber planer, it appears to be loafing along while the 6-10-A1 with 10-inch rolls is working much harder.

Narrow lumber planer described

Much of the lumber in the world today is made from small logs, 4 to 20 inches in diameter. Most of it is made into narrow lumber; 2x4s and other standard sizes of construction material. The Stetson-Ross narrow lumber planer has been designed to streamline the surfacing process in today's lumber operation.

Rough kiln-dried lumber is processed through a planing mill, as the final step in manufacture prior to shipping. Stickered packages from the kilns are brought to a tilt hoist and broken down for planing. In Crown Zellerbach's Armstrong, British Columbia, mill a random-length 2x4 line is beside a line on which only 8-foot studs are run.

Long chain sections enable lumber to accumulate ahead of a high-speed planer. A moisture dropout with a Sentry moisture detector monitors the incoming flow of lumber, rejecting any boards that are not dry. A Stetson-Ross lumber accelerator receives the transverse flow of lumber and forms it into a continuous

Doors, lined with sound-deadening material, protect planer operator from contact with moving machine parts. *Photo by Monroe & Akins*

Sound-deadening material covering sidehead hoods and top-head shaving spout reduces noise levels at the narrow lumber planer.

Photo by Monroe & Akins

ribbon ahead of the planer. These accelerators have hydraulically powered top rolls, which are air tensioned. Ten-inch table rolls start the lumber onto a live bridge, where it is butted.

The new 16-knife Stetson-Ross narrow lumber planer operates beside a 10-knife Stetson-Ross 6-10-A1. The narrow lumber planer runs between 750 and 850 feet per minute on random-length 2x4. Unlike other heavy-duty planers, this one has the top and bottom heads overhung. Tapered spindles are turned from 6-inch-diameter shafts. The cutting circle is 12½ inches.

Sideheads are arranged in the conventional manner, but the outside chipbreaker is air pressured. The top-head chipbreaker is also air tensioned for positive control of lumber at high feed rates. The 16-inch-diameter infeed and outfeed rolls are also air tensioned. One advantage of using air cylinders to tension both the chipbreakers and the feed rolls is that they are easily raised to clean out breakups.

Changing heads on this machine is very simple. The top head is removed in a manner very similar to sideheads on a conventional

Automatic stone advance for top- and bottom-head jointers eliminates need for operators' hands to enter the cutter area while jointing.

Photo by Monroe & Akins

planer-matcher. Both the top and bottom heads, as well as the sideheads, can be removed for servicing in the grindingroom because they mount on tapered spindles. A set of spare heads can be maintained in readiness for quick replacement during a normal run.

Sideheads are jointed with a detachable jointer. The top and bottom heads have built-in jointers. During the top-head jointing operation, the automatic ratchet advances for the jointer stone. This device allows the operator's hands to be kept well away from the cutterhead. On the operator's console, feed speeds can be adjusted to suit lumber conditions.

Lumber from the narrow lumber planer passes onto a grading table, where it is marked for trim and grade before going to a Stetson-Ross high-speed mechanical-lift trimmer. As lumber is loaded onto lugs, the trimmerman reads the grader's marks and punches in trim, grade, and sort. These trimmers operate between 50 and 70 lugs a minute, depending on the material being run. From the trimmer, lumber continues into a sorting system where it is stacked and packaged for shipping.

The Stetson-Ross narrow lumber planer is available in three sizes—4x4, 4x8, and 4x12—horsepowered for 8 up to 20 knives, with or without profilers. Ease of maintenance and simple design, together with heavy-duty feedworks and large cutting circles, make this machine well suited to today's modern planing mill for narrow lumber.

We have had success in marketing this new machine. Seven have been sold. Purchasers are Delah Timber Co., White City, Oregon; Diamond International, Passadumkeag, Maine; Pope & Talbot, Hudson, Ontario; Sooke Forest Products, Sooke, British Columbia; Weyerhaeuser Co., Lamar, Alabama; and Boise Cascade Corp., Kenora, Ontario.

The other film shown at the Clinic was taken at Biles Coleman Lumber Co., Omak, Washington. This planing mill is consistently the highest speed mill we know. It certainly is the most powerful 25-inch machine with 170 horsepower on the top cylinder to handle wide shop at high speed. It produces dimension, studs, shop and box—all at high speed.

Among our main concerns was how to grade the material at 100 lugs per minute, particularly shop. Our solution was to design a three-level grading system so that the grader sees only the lumber he is to grade and at a maximum speed of 33 lugs per minute.

After grading, the three lines of lumber merge for trimming and grade marking. Material then enters a high-speed sorting and dual stacking system supplied by Lumber Systems, Inc. Lumber tally is automatic as a package is made. A tally tag accompanies the load.

High-speed planing mill for mixed species

Biles-Coleman Lumber Co., in Omak, Washington, processes the output of three sawmills through one high-speed planing mill.

Rough lumber from the Omak sawmill is introduced to the planing mill with a kiln crib breakdown system, while lumber from sawmills at Grand Coulee and Twisp, Washington, are brought in over a tilting unloader. A variety of lumber products sawn from mixed eastern Washington softwood species are run through a system of machinery supplied by Stetson-Ross and Lumber Systems Inc. The equipment is designed for high production of 1-inch lumber, shop lumber, and 2-inch dimension. Over half a million board feet a day are surfaced, graded, trimmed, sorted, and packaged for shipping in this planing mill.

Material from the breakdown hoist is transferred to a Stetson-Ross Model 12 lumber accelerator. This feed table receives the transverse flow of lumber and forms it into a continuous ribbon ahead of a planer-matcher capable of feed rates up to 1,200 lineal feet a minute. Feed rolls on this unit are 12 inches in diameter. A powered bumper smooths the transfer of lumber onto the accelerator and underneath two overhead press rolls. Both the press rolls and the bumper are hydraulically powered and air tensioned. A 13-roll live bridge provides space for the lumber to butt between the accelerator and the planer.

An 8-inch by 25-inch, 20-knife, Stetson-Ross 6-14-D1 is designed to produce quality millwork at very high feed rates. Heavy-duty feedworks, powered with a variable-speed drive, have 14-inch diameter feed rolls, which are hydraulically tensioned for smooth feeding at these speeds. The hydraulics work to raise the rolls when the ribbon of lumber is broken and to bring them down again when lumber reenters the machine. Air-operated chipbreakers precede the top cylinder. Cutterheads on this planer-matcher have a 12 3/16-inch cutting circle and are powered by 85-horsepower motors. The top cylinder has two motors, or a total of 170 horsepower. Top and bottom profile heads can be used for splitting and pattern work.

From the planer, lumber emerges on an offbearing belt and transfers to a Stetson-Ross angle slowdown belt. This slows the material down and deposits it on an even lumber line on the camelback transfer preceding the grading system.

A clamp-off device forms the lumber into 4-foot-wide sheets before a tipple feeds it to one of three grading levels. Behind the clamp-off, the tipple feeds the top level and then moves to service the middle level. After lumber transfers to the middle, the tipple moves down and the bottom level is fed. The tipple then returns to the top position and the cycle begins again. The speedup belts under the clamp-off on the left and the belted tipple on the right are air actuated and controlled automatically. If any grading level becomes full, the tipple will bypass it and go on to the next position. Variable-speed drives are used throughout the system so timing can be adjusted to match planer output and suit changes of the material being run through the high-speed system.

Following lumber up the top level, it is fed to a lug loader. Here, individual pieces are metered onto a lugged chain for grading. The grading area is lighted with special lamps, and automatic board turners turn each piece for inspection on both sides. Each grading level has a single grader. Because the flow of lumber from the planer has been divided between the three levels, each man is concerned only with the lumber at his grading station. He is able to key in trim, grade, and sort signals to memory, without the distraction of seeing boards that have been graded by the other graders.

Back at the tipple again, lumber is fed to alternate grading levels. From the middle grading system, lumber goes up the incline to the top-level lug loader. The tipple then feeds a sheet to the middle level. All three grading stations have similar equipment systems.

Lumber from all three grading stations flows back onto a single lugged chain. The entire system is timed so each piece of lumber is deposited into a lug space that will convey it on through the trimming, grademarking, and sorting system. The electronic memory carries signals for each piece for trimming, grademarking, and sorting actuation downstream. Machinery is designed to handle up to 100 pieces a minute. At capacity, each grader will process a maximum of 33 pieces a minute. Here again, piece rate can be changed with variable-speed drives to suit particular lumber conditions.

From the top grading station, lumber feeds down the decline belts and through the high-speed Stetson-Ross trimmer. Behind the trimmer, it is marked for grade and continues up the sorter tower. Ahead of the trimmer, ending rolls move each piece of lumber into position for precision double-end trimming. Trim signals, carried in memory from the grading stations, selectively actuate Stetson-Ross "orbi-lift" mechanisms, which lower and raise the appropriate saws at up to 100 cycles a minute. These mechanical lifts are cam actuated for smooth operation and are fired electrically.

Behind the trimmer, ending rolls reestablish the lumber line as material passes to a Stetson-Ross end printer and a label grademarker. Signals still in memory from the grading stations actuate the labelers, which put a certified grade label and a product identification label on each piece. Paper grade labelling identifies each piece without defacing it as conventional ink stamps do.

Prior to going up the tower, boards can be diverted to a hand-pull chain underneath the sorting system. The pull chain dropout can be actuated by memory signals or for problem material on the chain.

The Lumber Systems sorter has a Z-shaped tower. Pieces still on lugs enter the base of the tower and start up a reverse incline, which extends back over the grademarking table. Lumber is carried face down. The Z-shaped tower is designed to conserve space while it transfers lumber to the top of the sorting system.

This sorter has twelve trays and has been designed so that two more can be added to the top. From this point, lumber travels down an incline until it is smoothly diverted into the appropriate tray. The front of each tray has a skatewheel accumulation section where the individual pieces of lumber are formed back into a solid sheet. Each tray is programmed for pieces with a certain width, thickness, and length—of a certain grade or predetermined mix of grades. Behind the skatewheel sections, the remainder of the trays are designed with powered chains, which are started and stopped to index sheets of lumber forward, making room for more material to accumulate.

Because lumber can be easily controlled as it travels downward, the transition made by the diverters at each tray is very smooth. Lumber Systems rotary diverters turn into position to bridge the gap between the lugged decline chains and the front of the skatewheel sections at each tray. The diverters are timed with the lugged chain and actuated by memory signals keyed in from the grading stations. Diverters have clutch mechanisms capable of cycling up to 100 times a minute.

Pieces of lumber are automatically counted as they divert to each tray. When the appropriate number of pieces has accumulated for a complete package, the powered trays convey the completed sheet away from the skatewheel accumulation sections. While the completed package is feeding down the tray to the stacking system, two sets of hook stops in the skatewheel section retard lumber until the tray has been cleared. The first set holds material until the back end of the skatewheel section is clear. It then releases it to the second set of stops. It is held here and continues to accumulate until the last board in the outgoing package clears the first tray chain section, leaving it free to begin receiving material for the next package. This Lumber Systems sorter is long enough to allow accumulation of one 2-foot by 4-foot package of 1-inch lumber in each tray or two packages of 2-inch dimension.

The outfeed end of the tray system is serviced by two Lumber Systems stackers. A tipple for each stacker receives completed sheets of lumber from the trays. The top tipple can service all twelve trays and the lower one is used for the bottom six plus the pull chain. Two stacking systems are required, because when the planing mill is running to capacity on wider material, trays are discharging a package a minute. The Lumber Systems reciprocating-arm stackers operate at 12 or 18 cycles a minute.

As lumber approaches the top stacker, adjustable course dividers automatically feed the right number of pieces over the stacking arms. The arms feed initial courses while tie-strip layers automatically position themselves for the length of lumber being stacked. Tie strips are discharged twice for each 2-foot-high package to hold it together as it transfers out of the system. Stacker arms are mounted on a carriage driven by a horizontal bell-crank. The arms automatically lift at the beginning of the cycle and lower to deposit lumber on the package by a cam mechanism powered from the bell-crank drive. This enables the stacker to cycle smoothly at different rates and for different sizes of material.

When a package is completed, the stacker lift raises so that accumulation arms can be retracted. The lift lowers the package to an outfeed rollcase and the accumulation arms move back into place to receive the first courses of the next package. Meanwhile the lower tipple empties one of the bottom trays and transfers the material to the bottom stacker. This unit is similar to the top stacker, with the same drive system and automatic tie-strip layers.

These stackers have controls so that they can be operated by one man when certain kinds of material are run—or by two operators on high-speed production of 2x8s. Indicator lights at the stackerman's consoles monitor the accumulation in each tray and show the operator when a tray is full.

Back at the stop stacker, the stacker lift cycles to remove a completed package. The accumulation arms retract and the load is lowered while the arms move back into place for the next package.

Out of the stackers, packages are strapped and readied for shipping. The grading, trimming, and sorting systems in this mill make it possible to produce efficiently quality lumber at high production rates.

Part two by R. Brian Parsons

Crown Zellerbach has been reorganizing its operations in the interior of British Columbia over the last several years. Part of the reorganization consisted of building a new mill to operate on small wood at our Armstrong, British Columbia, site. The mill was designed to produce random 2x4–a single product; however, some 2x3 and 1x4 are always generated as falldown from such an operation.

The output of this plant was to be about 315,000 feet per day, in two shifts. This quantity included about 70,000 feet of 8-foot 2x4 generated from plywood cores, which left 240,000 feet of random length 2x4, from 10 to 20 feet, to be dressed daily. The arithmetic of these quantities indicates a sustained planer speed somewhere in the order of 800 feet per minute.

We had on hand at another plant, a Stetson-Ross 6-10-A1 planer, which appeared to be the right machine for studs. This left us with a need for a high-speed machine with 16-knife heads, suitable for kiln-dried stock. When the market was examined, we were offered the standard heavy-duty 6-inch by 15-inch high-speed planer that manufacturers have been producing in quantity for years. There was no response to inquiries for a small machine.

It would have been wrong to purchase a machine capable of handling 6-inch by 15-inch material to plane 2x4s. Probably many other companies find themselves in a similar position. Our views on the situation were made known to Stetson-Ross, and after some hesitation, they suggested a concept which they had under development for a small, high-speed machine. The machine described met our requirements for both size and speed, and it was priced considerably below the cost of the large heavy-duty high-speed machines.

On examination, it appeared that the principles on which this narrow lumber planer was based had been used traditionally, and were sound. The only departure was that the frame would be a stress-relieved weldment, rather than the castings used in the past.

I had some confidence in Stetson-Ross, since they have been in the planer-building business for a long time, and it seems they intend to stay in business for some time to come. Their operations were close at hand, so they could be easily reached—an important item when one is engaged in the design and construction of plants.

The machine was built, installed in our plant, and has been running since the end of 1973. I feel we made the right decision in the design specification, and purchase of the planer. This has been reinforced by the rapidity with which others have placed orders for the same machine. The demand probably existed for some time, waiting for the product to meet it.

18

Save wood/raise grade with abrasive planing

Sherman E. Kirchmeier, Oroville Division Manager, Louisiana-Pacific Corp., Red Bluff, California

Abrasive planing is being discussed more and more in our industry. Because of the outside pressures being applied to our industry, wood utilization is becoming synonymous with survival. It is no longer a subject for future consideration.

Precision sawing is already a reality to most of us. Certainly we must examine any system that will complement this program or enhance profitability. Abrasive planing could be one of these systems. Let us examine its advantages and its current status in our industry.

There are two types of systems available. They are known as the *batch system* and the *in-line system.*

Batch system

The batch system is the type Louisiana-Pacific Corp. has installed at its plant in Oroville, California. It involves a dual line including a pair of two-headed edge planers, two accumulating lines, and a four-head Timesavers abrasive planer. The planer is 52 inches wide, similar to a plywood or particleboard sander.

The system operates thus: Boards are first passed one at a time through the edge planer. Next they fall into the accumulators, which align the pieces for entry into the wide-belt machine. The machine sands up to thirteen 2x4s at one time. If stock removal is not excessive, the wide-belt abrasive planer can be fed at a minimum of 125 lineal feet per minute, or 1,625 board feet per minute, which is a speed certainly advantageous to our industry.

In-line system

The in-line system is the type installed at our Samoa, California, plant. It is just what the name denotes. It entails abrasive planing on the top, bottom, and two sides, one board at a time. It provides the flexibility to mix profiling knives and saws with the abrasive heads. Thus, it would be the natural system for easiest conversion of existing planing mills. It permits installation of the abrasive planer in the area where the knife planer now sits. This uses no more room and may permit use of existing offbearing equipment. Where space is an issue or established plants plan to renovate their surfacing department, the in-line system would be the likely choice.

Let us examine each system on its merits and list the advantages and disadvantages of abrasive planing itself.

If the mill requires a consistent, heavy cut with high speeds, there is very little economic gain with in-line abrasive planing. If, however, a plant is sincere in precision cutting and/or avoiding downgrade due to chipping out or tearing grain, then this is an ideal system.

This discussion should not be taken to suggest that all abrasive planers cannot remove heavy stock, or that they cannot achieve both good production rates and grade recovery. Northwest Hardwoods Co. at Arlington, Washington, has shown that this can be done. Since late 1972, this plant has been abrasive planing its alder and bigleaf maples using a batch system. Due to the nature of these species, this system constantly fights thick stock and distorted boards. But its abrasive planer handles them with little trouble. I understand that belt costs are in line. Production is reported excellent and grade recovery is better than had been anticipated. This company's management must believe in this system since they will soon install another abrasive planer at their Centralia, Washington, plant and they have another on order.

A board passing through in-line abrasive planing operation at Louisiana-Pacific Corp.'s Samoa, California, mill.

Low feed rates best

It is axiomatic that the less stock given the abrasive planer, the more it excels. Feed speeds increase, abrasive belt life increases, and the grade recovery remains the same. At our Oroville plant we can shoot for much closer cuts with no fear of losing grade. We have often taken dried lumber that had only 1/64 inch total to clean and actually were able to burnish the board enough to maintain its grade. This has been especially lucrative in our vertical grain shop grades.

It should be emphasized that shop lumber which contains large knots is never chipped or torn; this enables us to pick up even more grade. To gain two grades is not uncommon, which could not be done with a knife planer.

When we first started this plant and our thickness variation was substantial, we had only one accumulator loading our Timesavers abrasive planer. It could handle the batch of boards at only 80 feet per minute. Our finish, grade, and belt life did live up to our expectations, but the total footage output did not. Those slow

Samoa in-line system permits mixing profiling knives and saws with abrasive heads. Much of mill's production is in shop grades.

At Louisiana-Pacific Corp's Oroville, California, plant, a batch accumulates in foreground as operator peavies boards in batch at rear to assure smooth entry onto feed line.

Control panel for the batch planer at the Oroville plant.

feed speeds made it necessary for us to run the finish department six days, three shifts. Since then, however, we have learned more ways to control our thicknesses by better sawing and drying techniques. On our kiln-dried lumber, we set our gang-edger saw at exactly 1 21/32 inches, running at 300 feet per minute. Yet we consistently attain high accuracy throughout the length of the boards. Our consistency is 98 to 100 percent, plus or minus 1/32 inch, believe it or not.

We have also learned enough about our dry kilns to achieve higher efficiency. We can now easily control the moisture content without getting any more than the allowable skips.

Production doubles

As a result, the abrasive planer's production has doubled. In fact, we have since added another accumulator, including another set of side heads, and are running the finish department five days, two shifts.

The side-head knife planers have the advantage of maintaining speeds of 750 feet per minute or faster, while the abrasive planer handles this production at feed rates of 150 feet per minute or better. This plant has run as much as 321,000 board feet through the line in nine hours. It consistently runs 240,000 board feet every eight hours. The mill does this with fourteen men per shift, including three graders. Many mill men may question whether three graders can adequately handle this amount of wood. I assure you that as long as it is brought to him properly, one man can grade 100,000 board feet a shift. We have also learned we could eliminate four more men if we could make more modifications.

Belt costs have been reasonable. In 1972 they ran between 50 and 75 cents per thousand board feet. In 1973, we reduced them to between 25 and 50 cents and they have been as low as 14 cents per thousand. In 1973 we reduced our total surfacing department costs by about 15 percent, as compared with 1972. Costs now range in the lower 10 percent of all our sawmills. This figure is low when you consider that our volume per eight hours is, in most cases, two to three times larger than our other mills.

Then there is our ability to recover virtually 100 percent of our vertical grain shop grade lumber. Consider again the high production and large wood recovery on construction grades; consequently, total surfacing costs are very economical.

Outfeed of the Oroville abrasive planer. Painters reportedly prefer sanded surface to one from traditional knife planer.

Advantages noted

Abrasive planing has other advantages: abrasive belts are easy to change and have no maintenance costs, which knives do. Short stock, 8 feet and less, is easily fed through both the side-head planers and the batch-type abrasive planer without the typical breakup problems. Crooked lumber can also be handled easily. We can hold tolerances in thousandths of an inch from side to side and from end to end without any skips. This precision is made possible by the unique feeding systems, pressure plates, and fixed heads which this abrasive planer has. All of these factors have, of course, reduced downtime as compared with a knife planer.

Abrasives improve

We are using conventional open-coat type abrasives in 24 and 40 grit. The 3M Co. has been supplying us with newer, more durable materials which have cut our costs. However, the 3M Co. is working on still newer products to reduce belt costs further. The new abrasives should lend themselves to even more versatility.

We are hoping that these new products will allow us to run our small amount of green lumber completely wet. Thus far, we have had to dry the perimeter of our green lumber before passing it through the abrasive planer. Though this has not been the great disadvantage we first thought, we are still striving to enable the abrasive planer to run completely wet lumber. The 3M Co. has been working on this problem and hopes to have such an abrasive product soon.

We are certainly pleased with this system. It has more than lived up to our expectations and allowed us to utilize our sawing techniques. If anyone is serious about cutting lumber accurately, we recommend abrasive planing.

Samoa plant

I would like to update the information on our system at Samoa, California. It is the first in-line abrasive planer in our industry. Its main purpose is to dimension and finish finger-jointed stock. This abrasive planer was manufactured by Kimwood Corp., of Cottage Grove, Oregon. It includes a molder-type feed system, two top and two bottom 15-inch-wide abrasive heads. It is followed by a knife molder, which is connected to the sander. This molder, which can use our existing profiling heads, is necessary in making many of our type of products, including facia boards, S4S, and others. The primary purpose of the machine was to eliminate chipping out on finger-jointed stock and tearing grain in architectural boards. Prior to the installation of the planer, this mill was downgrading approximately 40 percent of its glued stock to B grade. This was caused by knife planers chipping out the offset glue joints. The abrasive planer, as planned, has completely eliminated this problem. As a result, our recovery is 100 percent.

This system had many imperfections that none of those involved realized initially. Kimwood Corp. has made some major on-the-spot modifications. The 3M Co. also learned quickly that its conventional abrasive materials were far from adequate. The firm had to fabricate new types of belts that, up to then, no market had ever seen. With everyone's effort and cooperation, this system is running satisfactorily. Belt costs, which started high, are now very reasonable. Belt costs initially ran about $1.70 per thousand board feet. The past few months they have run between $0.25 and $0.30 per thousand. As you also would expect with these high costs, in the beginning production was poor. As modifications were made both to the machine and to the abrasive belts, this plant's production continued to improve. While the abrasive belt costs were improving, so was production. Since October 1973, this line has been averaging 35,000 to 45,000 board feet when running 1x8 and 1x6 material. Note that footage increases as board width increases because the wider boards can be run at the same feed speeds as the narrower boards.

3M develops belt

With a new belt developed by the 3M Co., Louisiana-Pacific increased production another 33 percent (during February 1974) while reducing belt costs another 80 percent. This may be hard to believe, but during this recent test, we ran up to 60,000 board feet per shift of 1x6 and 1x8 material back to back with no belt changes. This was more than 100,000 lineal feet of material in eight hours. One morning we ran 35,200 board feet of 1x10 stock through these abrasive belts in less than three hours. We removed 1/8 inch total—at 400 feet per minute. We broke no belts; we had absolutely no joint damage or grain tear. We are very pleased to attain that kind of volume with low costs—and while recovering the grade we originally set out to save. It has been a difficult road for everyone involved but the end results have justified it.

We are seriously considering adding more heads to add more versatility to this system. We do not use precision sawing at this plant. But we do believe that we could still run all types of dried redwood through this abrasive machine very economically if we had more heads. The benefits, we feel, would soon be apparent.

System still young

Abrasive planing is certainly in its infancy. There are still problems to be resolved. For example, our biggest problem is that, while we are experimenting, we still have to produce. We do not have an abrasive planer that is set off to one side for experiments.

We are now working on changing belt speeds and changing the feed. An abrasive belt is just like a saw. When you overload a saw, you fill the gullets, creating problems. This is true with the sander. Since we can only remove so much material, we must increase the speeds. I hope to have more factual information on both of these problems by the end of 1975.

In the meantime, this subject is so active that changes are occurring almost daily. New factors are being introduced, in terms of equipment and products, that will cause important changes. As these improvements are made, we as an industry will be the direct beneficiary. Everyone in the sawmill industry agrees that wood-saving systems are becoming more essential. Louisiana-Pacific Corp. obviously believes that abrasive planing is one of those systems. I recommend that everyone in the sawmill industry explore the possibility of abrasive planing. It could be the answer.

19

CRT drying for reduction of degrade

Dr. Dallas S. Dedrick, Lumber Drying Consultant,
Irvington-Moore, Portland, Oregon

When I discuss the lumber drying technology commonly known as CRT, or continuously rising temperature, I usually emphasize the theory behind it. A procedure which differs markedly from accepted technology requires explanation and justification.

Mill operators, however, are much less interested in the 'how' and 'why' than in what it costs and how well it works. So I want to deal first with the performance records of CRT, both in the plant and in the laboratory. Then, I will answer a question often asked by mill operators: How desirable is remodeling an existing kiln to use CRT technology, compared with replacing that kiln with a new facility engineered to provide the features which have been identified as most effective?

CRT process explained

CRT is the commonly applied designation of a novel, patented process of controlled heat transfer from a heat source to wood in a dry kiln. CRT is not a schedule. Rather, it is a process in which energy is transferred from air to wood at a substantially constant rate, resulting in moisture removal at a substantially constant rate.

Conventional methods of drying invariably result in falling-rate drying—the rate at the beginning of the drying period is much greater than the rate at the end. If a falling rate of drying can be replaced by a constant rate, kiln residence time can be substantially reduced.

Although reduction of kiln residence time may appear important at first glance, its importance is minor compared with another factor associated with drying—economics.

Assume an extreme example. The operator of plant A uses a kiln which costs $100,000 and holds 84,000 board feet of lumber. His drying schedule is 48 hours. In an 8,400-hour year he can dry 175 charges, or 14.7 million board feet. If 20 years is his chosen amortization period, the kiln throughput will be 294 million board feet. The capital cost of his kiln is therefore $0.34 per thousand board feet.

The operator of plant B dries the same amount of lumber but uses a slow schedule requiring the use of three kilns. The amortized kiln cost is obviously three times higher than in the first example, or $1.02 per thousand board feet. The difference is $0.68 per thousand—certainly not an inconsiderable amount.

Product value raised

Studies have shown, however, that product values of kiln charges may represent differences on the order of 10 to 20 times the number in the extreme example. Thus, if the product value of the more slowly dried material were $8 per thousand board feet higher than that of the faster dried material, the operator of the slower process would be $7.32 per thousand ahead, despite the triple capital investment in kilns.

This example serves to stress a key fact: product value is a more important criterion than kiln capital cost in the selection of a kiln or a drying process. CRT, however, both reduces kiln residence time and yields high product value. Thus capital cost is reduced, because fewer kilns are required, and a higher product value results also.

In the three years since its introduction, the CRT process has become widely accepted. Approximately 3 percent of the softwood lumber dried in the continental United States and Canada is now processed by this new technology. CRT has been accepted by Weyerhaeuser Co. Its return is already reported to be approximately 10 times the $1.4 million spent on development of the constantly rising temperature drying technology.

Field results shown

Just how well is the technology performing in the field?

In routine commercial operations, drying times, although much shorter than those normally used, are not necessarily consistent for the same items (Table 19.1). Kiln operators have chosen to use slightly different rates of energy transfer for varied reasons. No one knows, yet, how fast the drying rate for a given lumber item must be to achieve maximum product value.

We now are accustomed to people reacting with something akin to awe and wonderment when they see the first CRT-dried kiln load discharged. At a new CRT installation, people universally mention the unexpected flatness and light color of the lumber.

Output quality praised

The quality is described in the words of responsible and knowledgeable plant personnel in qualitative rather than quantitative terms. To obtain reliable numbers for sound comparisons, a controlled, large-scale study, involving carefully matched samples, should be done. For statistical validity, at least a crib of lumber should be used for each comparison. The lumber should come from the same logs, preferably from adjacent positions in the logs, to insure comparable "dryability." Each board must be followed through the processing–from kiln loading to the dry, surfaced, and graded piles–and the product value per thousand board feet charged should be calculated. The process should then be repeated until the computer indicates that one schedule or kiln or process is consistently and significantly better than the other within a calculated range of values.

This is a slow, tedious, and costly process. To my knowledge, it has actually be carried out for only two lumber items–Douglas fir dimension and southern yellow pine dimension. In both cases CRT processing yielded statistically valid product values several dollars

Table 19.1. Data from Mills Using CRT Drying Process—by Species

Company	Facility	Moisture content change	Drying time	Quality
		Douglas Fir Dimension		
F	Remodel–Steam	45%—13%	40 hrs.	Improved
B	New–NG	50%—13%	24 hrs.	Excellent
D	Remodel–Steam	60%—15%	40 hrs.	Excellent
E	Remodel–Steam	55%—15%	38 hrs.	Excellent
H	Remodel–Steam	60%—14%	34 hrs.	Improved
		Western Hemlock Dimension		
A	New–NG	85%—17%	30-32 hrs.	Excellent
B	New–NG-O	90%—17%	30 hrs.	Excellent
C	Remodel–Steam	85%—15%	48 hrs.	Improved
D	Remodel–Steam	90%—15%	45 hrs.	Excellent
E	Remodel–Steam	80%—16%	48 hrs.	Improved
		Canadian Spruce – Jack Pine Dimension		
I	New–NG	60%—16%	24 hrs.	Excellent
J	New–O	75%—19%	26 hrs.	Excellent
K	New–G-O	70%—15%	26 hrs.	Improved
		Southern Yellow Pine Dimension		
L	New–Steam	120%—16%	23 hrs.	Min. crook and warp
M	New–Steam	110%—15%	24 hrs.	Excellent
N	New–Steam	115%—15%	23 hrs.	Excellent

per thousand board feet higher than either the conventional or the high-temperature processes in the comparison.

Because of the complexity and cost of the required testing procedure, few mill operators know just how much their kiln-induced degrade is costing them. Therefore, the quality is described in qualitative terms.

Experience with CRT cited

During the startup of CRT processing of Canadian spruce and jack pine mixtures, two significant incidents took place. In one plant, an official grader was heard to say that after seeing the CRT-dried lumber, he hoped he would never again have to handle either hi-temp or conventionally dried lumber.

In the second plant, a foreman sitting in his office some distance from the unstacker said, "Well, they've finished the CRT load and are starting on the regular material. Notice how the sound just changed from a series of soft thuds to sharp metallic sounds as the boards fall."

A good question is whether CRT can be applied effectively to all lumber items green from the saw. I don't know. There are many lumber items which have not been tested. Some experimental results are shown in Table 19.2. Recently I was carrying out experimental CRT tests on green, frozen, 9/4-inch, random width, untreated sugar pine shop grade lumber having a green moisture content of 215 percent. The fact that the lumber came out of the kiln straight, flat, and unchecked was completely overshadowed by the most severe brown stain I have ever seen. So it will take further testing before we know whether this item can be dried green from the saw in a short time without stain-inhibiting treatment.

CRT technology is being applied commercially to yield both

Table 19.2. Experimental CRT Drying Data

Item	Moisture content change	Drying time	Quality
6/4 Ponderosa Pine Shop	130%—6%	28 hrs.	white, flat, straight, no brown stain
Lodgepole Pine Studs	65%—15%	16 hrs.	light, straight
9/4 Sugar Pine Shop (untreated)	215%—15%	148 hrs.	flat, straight, no checks, severe brown stain
4/4 Red Maple	65%—6%	84 hrs.	excellent
6/4 Hickory	60%—6%	72 hrs.	excellent

short drying times and high quality in the processing of Douglas fir, hemlock, Engelmann spruce, jack pine, balsam, ponderosa pine, southern yellow pine, and furniture grades of birch and maple. In probably no case, however, has the full CRT potential been realized.

Reference to the experimental investigation of lodgepole pine studs in 16 hours with apparent good quality (Table 19.2) illustrates the point. Perhaps an 8-hour schedule with minimum degrade is possible.

Research continues at both Irvington-Moore and Weyerhaeuser to optimize drying economics further by learning still more about the mechanism of lumber drying and to establish applicability of CRT to other lumber items.

Remodel—or buy a new kiln?

It has been repeatedly said that CRT is not a schedule but a process of controlled heat transfer yielding a substantially constant rate of moisture loss with inherent protection against kiln-induced degrade. The elements of CRT are the temperature control equipment and humidity control techniques—very much like a conventional kiln.

The dry-bulb control equipment must provide the predetermined program of dry-bulb temperature rise starting at the lowest possible level while still drying completely across the load, and ending at whatever temperature the kiln can produce or the wood withstand without degrade. Fast drying rates obviously require more available heat transfer than do slow rates.

The humidity control system involves the controlled intake of dilution air, since the mixing of dry outside air with spent air from the kiln is how the wet-bulb temperature depression is established. Fast drying requires a large wet-bulb temperature depression, which can be maintained by mixing a relatively small quantity of dilution air with the spent air at higher dry-bulb temperatures. However, at lower dry-bulb temperatures a low humidity can be reestablished only by using much more outside air. The typical conventional dry kiln has an intake venting capability of only 5 to 10 percent dilution air. Therefore, the dry-bulb temperature must be in the 150- to 160-degree Fahrenheit range to achieve the

40-degree wet-bulb depression necessary to dry reasonably fast with this relative quantity of dilution air. If it is desired to start drying at, for example, 120 degrees Fahrenheit, more dilution air is required. If drying is started at 100 degrees Fahrenheit, still more outside air must be available.

Most CRT installations provide for more than the conventional quantity of intake air by using either larger-than-normal vents or controllable fans, which force the required amount of dilution air into the box.

Economics of kiln remodeling

Since the typical conventional kiln consists of a heating system and a controllable means of admitting intake air to adjust humidity, CRT technology, in theory, may be applied in any existing kiln. It is certainly true that best results are achieved by using a new facility in which all components are specifically engineered to operate together with maximum efficiency. Economic advantages are possible, however, by modifying existing equipment. In fact, most current CRT processing is done in remodeled facilities.

The extent of remodeling required depends, obviously, upon the characteristics of the existing kiln and the performance desired after remodeling. A kiln with large heat transfer capability and large intake vents, for example, might require minimal change to increase throughput by 10 to 25 percent. This might involve merely providing a dry-bulb temperature control instrument, modernizing baffling, and adequately instructing the operator. For higher performance, however, extensive changes in heating, air intake, and air delivery components may be required.

Costs of kiln modification

Some kilns which routinely operate on CRT have been modified at an equipment cost of less than $3,000 each. Another plant has remodeled a large battery of kilns, which were built in 1929, at a cost of approximately $30,000 per box. Still another plant is changing one-third of its old boxes at a cost of $80,000 per unit. This firm plans to retire the remaining boxes or to use them for standby or special purposes.

The decision to remodel or to install new kilns is entirely a matter of economics. Consideration must be given to (1) the amount of improvement in performance required; (2) the heating, venting, air circulation, and baffling characteristics of the existing box; (3) the state of repair of the existing box; (4) the capability of the existing box to accommodate the indicated changes (For example, if the length or the diameter of the fin pipe must be doubled, is there room for it?); (5) the cost of the changes; and (6) the anticipated low-maintenance life of the facility after modification is completed.

Irvington-Moore's people are prepared to evaluate existing facilities and estimate the performances to be expected from various amounts of modification. In the final analysis, cost will be the determining factor.

20

Ultrasonics to detect wood's defects

Erwin H. Bulgrin, Project Leader, Quality and Yield Improvement in Wood Processing, U.S. Forest Products Laboratory, Madison, Wisconsin

Will we see with sound to saw? Despite the catchy wording, the idea is not new. More than 300 years ago William Shakespeare said through King Lear, "A man may see how this world goes with no eyes. Look with thine ears." My colleagues and I believe you can see with sound. Before I explain how it can be done, it may be well to spend a few minutes discussing why we want to see with anything other than the two eyes we have.

Substitution problem

The relative importance of timber products in the United States economy has changed greatly over time as other raw materials, such as steel, concrete, aluminum, and plastics, have replaced or supplemented wood in various uses. With a strained raw material supply situation, and with increased prices of stumpage, equipment, and labor, prices of timber products can be expected to increase. So such substitution can be expected to continue and possibly increase. While the alternative of substitution may be one solution to the problem, major developments of the past year have made it painfully clear that it is not a desirable, long-run solution.

In this country, more than 35 percent of our iron ore requirement is imported, 90 percent of our bauxite for aluminum is imported, and an unclear percentage of our petroleum is imported. However, several things are eminently clear in this situation. First, increased dependence upon foreign sources for raw materials places our economy in a very tenuous position. Second, of all the major materials we require, our wood resource is the only one that has the potential of meeting our needs for the foreseeable future from native sources. Of crucial importance, this reliance on domestic timber resources can be achieved without the additional energy requirements and undesirable environmental side effects that accompany the exploitation of lower grade ores.

Wise use needed

Obviously, it behooves us as a nation to use our timber resource as efficiently and wisely as possible. We are trying to learn how we can improve our utilization efficiency. What we are learning in all quarters is that the key to the situation is the workman at the saw. The thousands upon thousands of decisions he makes each day determine the degree of efficiency at which we convert raw material into product, or put in another way, determine whether we operate at a profit or at a loss. These decisions are based, in large part, on what the workman sees, or thinks he sees, and his evaluation of this information. Believe me, he needs help! The inaccuracies in his information are reducing our conversion efficiency and our profit. *That* is why we are trying to see with something other than our eyes.

A number of systems or techniques can be used to perform the inspection function of the eye. In our original appraisal and screening, we looked at a variety of techniques and found they fell into several general categories: (1) those that evaluate density and variations in it; (2) those that evaluate color or light reflectance; and (3) those that evaluate configuration or pattern and variations in it. With techniques of this type, it is necessary to establish a norm, and base evaluation on deviations from it.

In other words, densities within this range are acceptable; densities above or below this range are not. These colors are acceptable; those are not. These configurations or patterns are acceptable;

those are not. Wood is a variable material. Consequently, it would be extremely difficult to establish these norms, and they would have to be monitored constantly to adjust for within-piece variation, between-piece variation, between-condition variation (green or dry), species variation, and so forth. Of the techniques we screened, ultrasonics—the technology of high-frequency sound above the range of the human ear—showed the greatest promise.

Cell wall soundingboard

Sound is transmitted through, or propagates through, wood in the cell wall. The density or specific gravity of the cell-wall material is about 1.5, and is the same for all species—both hard-

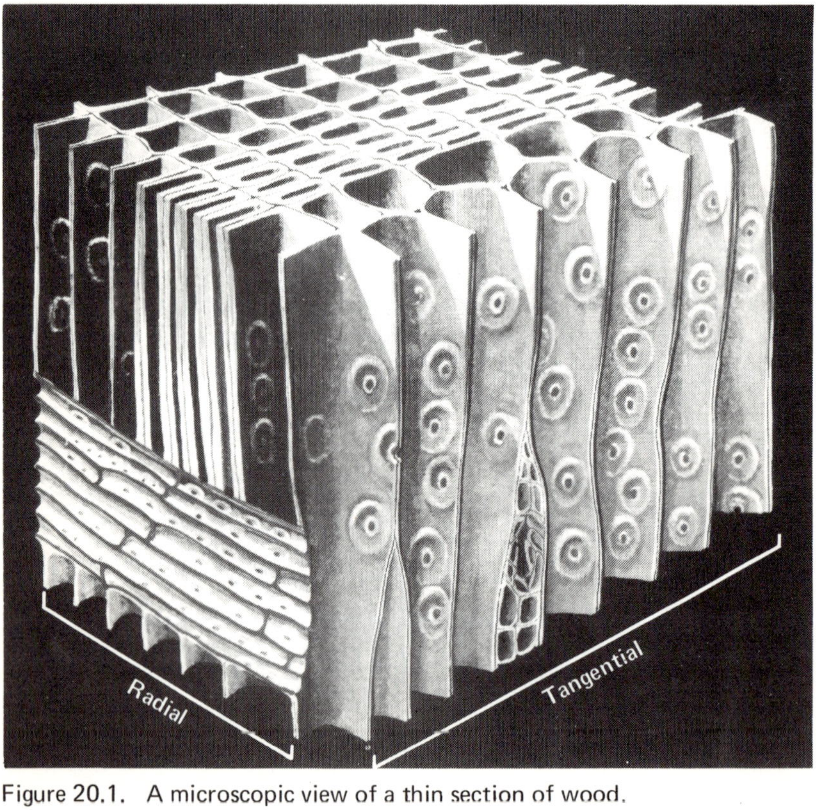

Figure 20.1. A microscopic view of a thin section of wood.

woods and softwoods. Consequently, the movement of sound through wood is not affected by the bulk density of a given volume of wood, which actually reflects only the amount of cell-wall material and void that is packed into a piece of wood of a given size.

Figure 20.1 shows a diagram of a microscopic view of a thin section of wood. As we see it, the material is oriented as it stands in the tree. Therefore, the vertical direction illustrates the vertical arrangement of cellular structure in the tree. The upper face illustrates a cross-sectional or log-end view of the material. The left face illustrates a radial or vertical-grain surface, and the right face illustrates a tangential or flat-grain surface. Sound, propagating in the cell wall, transmits fastest parallel to the grain, or in a vertical direction in the figure. It transmits slightly slower in a radial direction, and again slightly slower in a tangential direction. The reason for the differences in transfer time is apparent if one examines the cell-wall alignment. In the direction parallel to the grain, the cell walls are in an almost continuous straight line and sound can transmit rapidly along the straight path. In the radial direction (top to bottom), as illustrated in Figure 20.2, which shows a photomicrograph of a cross-sectional view, cell-wall alignment varies somewhat and the path is not as straight. In the tangential direction, left to right in the photomicrograph, the cell-wall alignment is poor and the path wanders.

Moisture no factor

Since sound moves in the cell wall, is independent of moisture content and bulk density, and is sensitive to grain direction, it forms the backbone of our system to sense and locate defects in wood.

Figure 20.3 is a photograph of our research equipment. I would like to emphasize that this is research equipment and not a production model. The equipment consists of the defectoscope, across the front of the picture; the electronic equipment, mounted in the rack, consisting of a time-interval counter, an oscilloscope, a transient recorder, a pulse generator, and the two controllers for the indexing system; a computer; a magnetic-tape drive; and a line plotter.

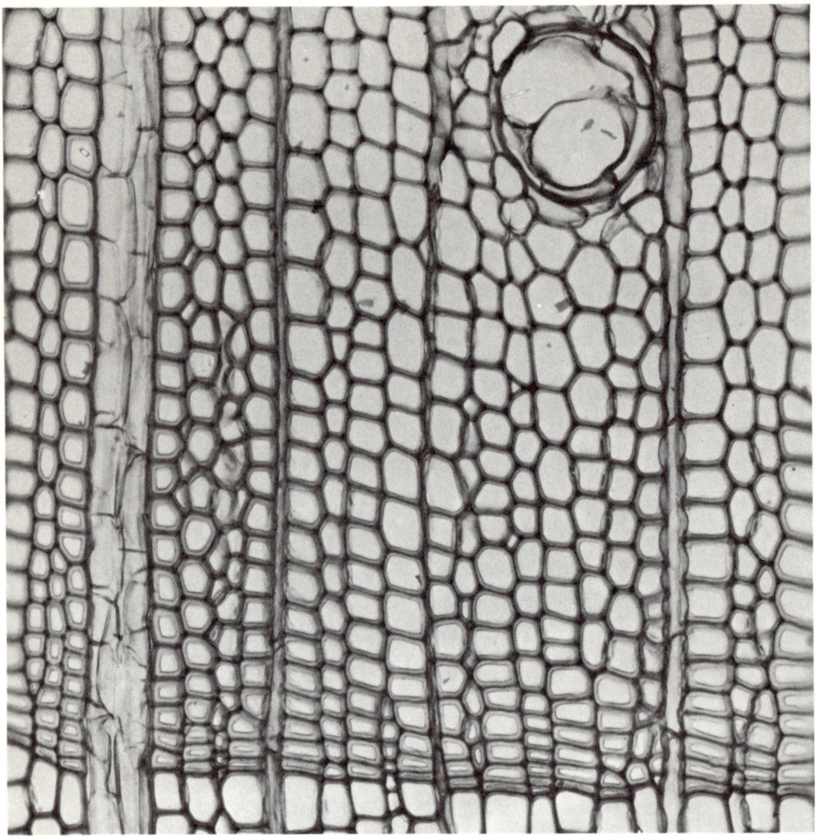

Figure 20.2. Photomicroscopic view of a cross section of wood. Horizontal cell walls are irregular in alignment, in contrast to vertical walls.

This is how the system works. A board is mounted in the defectoscope between the two transducers that are mounted on the lower end of the two blades you see rising above the bridge-like assembly. The computer then instructs the pulse generator to pulse the transducer crystal. It also instructs the bridge assembly to transport the transducers the full length of the board, then move up 1/2 inch, return the full length of the board, move up 1/2 inch, and repeat until the entire board has been inspected. At the same time, the computer instructs the time-interval counter to measure and record the transit time of the ultrasonic pulse at every 1/2-inch point along every traverse of the board. Transit

Figure 20.3. Sound measurement equipment in the U.S. Forest Products Laboratory, Madison, Wisconsin, is used to develop methods for detecting defects in wood. Water acts as couplant between transducer and wood.

time is the elapsed time starting when the sending transducer is pulsed and continuing until the signal passes through the board and is picked up by the receiving transducer. Typical transit times are around 60 microseconds, which is sixty-millionths of one second. This time count, in microseconds, along with location information of the reading point, is conveyed to the computer. The computer then digests this information and directs the line plotter to sketch what the system can see with sound.

Transit times traced

Figure 20.4 illustrates a contour plot of ultrasonic transit time. The computer, using a contour plotting program we borrowed from our civil engineering friends, has plotted all the transit-time readings as received from the counter. Several things are apparent in this plot. The ultrasonic transit-time contour lines duplicate to a high degree the general grain pattern visible on the surface of the board. Knots are obvious, and you will recall I said transit time is related to grain direction. This is also obvious in the deflected

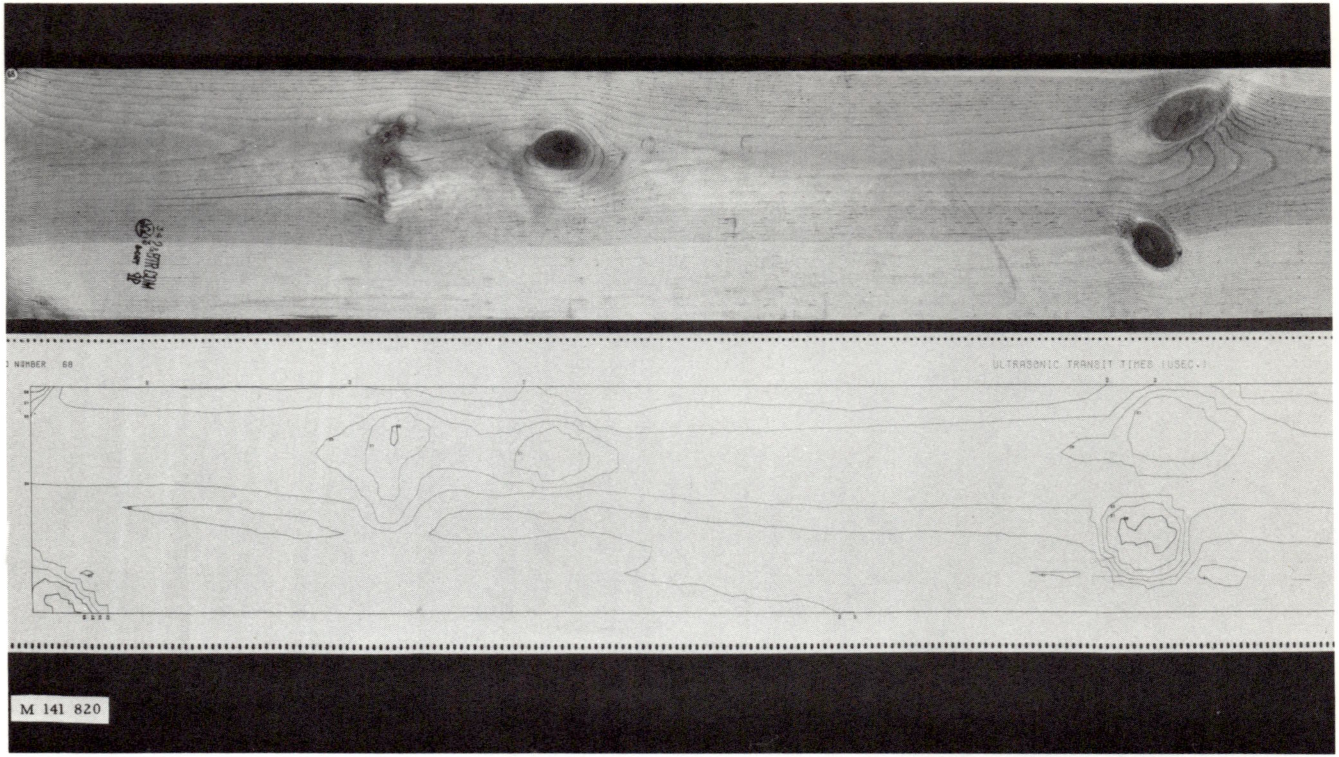

Figure 20.4. Contour plot of ultrasonic transit time (at bottom) reflects general grain pattern visible on board. Computer traces transit time readings, as received from counter, of ultrasonic impulses which bombard board.

grain surrounding the knots. I would like to point out one more thing. This system sees the board as transparent and integrates and displays the total defect area involved on both surfaces of the board.

It is interesting and fun to watch the plotter pen squiggle in those lines. But what can we do with a plot of this sort? Obviously we can get a good picture of the grain pattern, and areas where steep grain intersects the edge of the piece stand out. Steep grain at the edge may be related to strength, and this may be valuable information to the man who is attempting to evaluate the strength of the piece. At this point, we have no evidence to confirm this hypothesis, but we are investigating it.

In Figure 20.5 a plot of the same board is illustrated. However, the individual defect points are plotted and the knots have definite borders. The computer is now differentiating between defective and nondefective material, acceptable and unacceptable material. This is how the computer goes about making the decision. As the computer receives the transit time reading at each data point, it compares that reading with readings obtained from adjacent points. If the difference equals or exceeds a specified difference—in this case 0.50 microsecond—the data point is recorded as a defect point.

Furniture market

In looking at this plot, it is apparent that this system is potentially useful in a furniture cutup plant or a molding and millwork plant. The next step would be to develop a computer program that can locate the clear cuttings you require. Figure 20.6 illustrates a simplified version of just such a program. The legend at the top of the plot is defined as follows: S = the scale of the plot; L = length of the piece in inches; W = the width of the piece in inches; G = grid, or the distance between data points; D = difference in microseconds between data-point readings that indicate defect; SM = shortest acceptable length cutting in inches; KF = width of the saw kerf in inches. This is a simplified version of the program; only widths of 2 and 2 1/2 inches are specified and a minimum length of 6 inches is specified. To be of practical use, a program that can seek out a mix of widths and lengths and maximize yield is

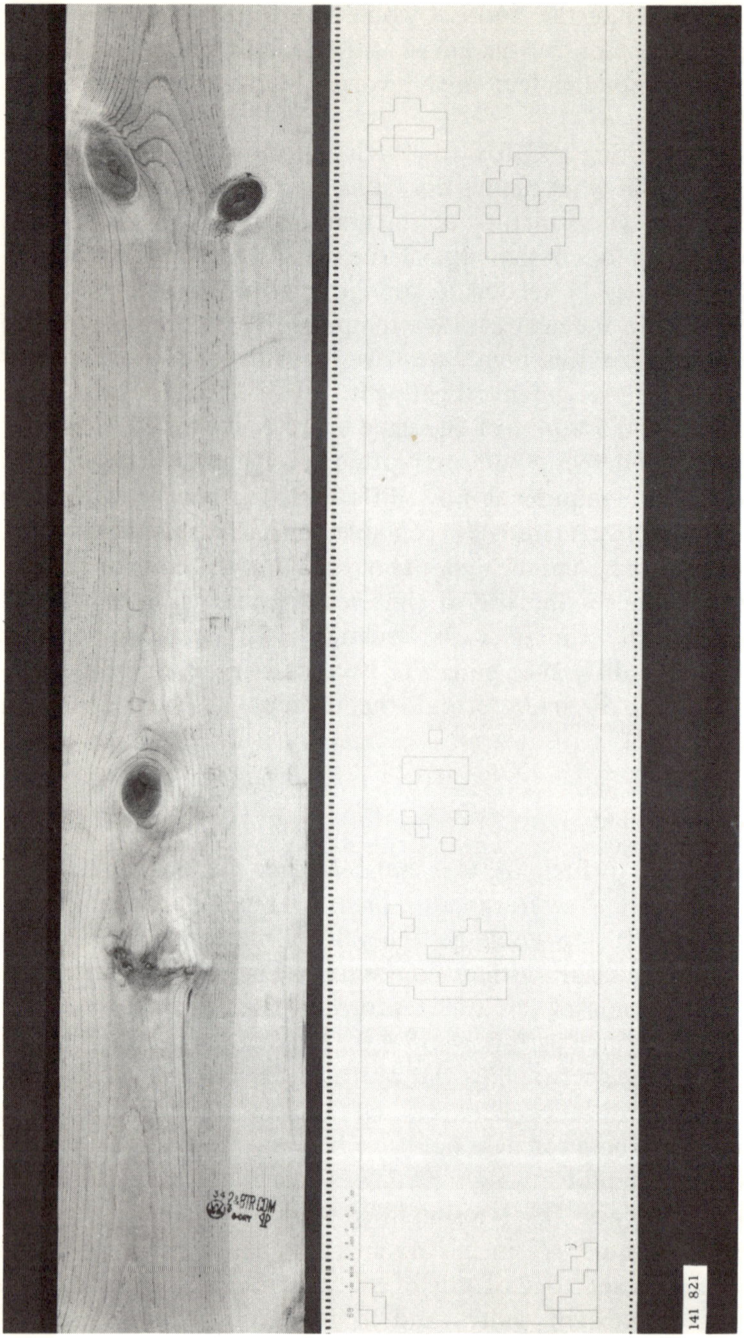

Figure 20.5. Computer pinpoints, and plots, defects by noting differences in transit time. Computer compares readings at adjacent data points and records as a defect any point showing variation equal to or exceeding that specified.

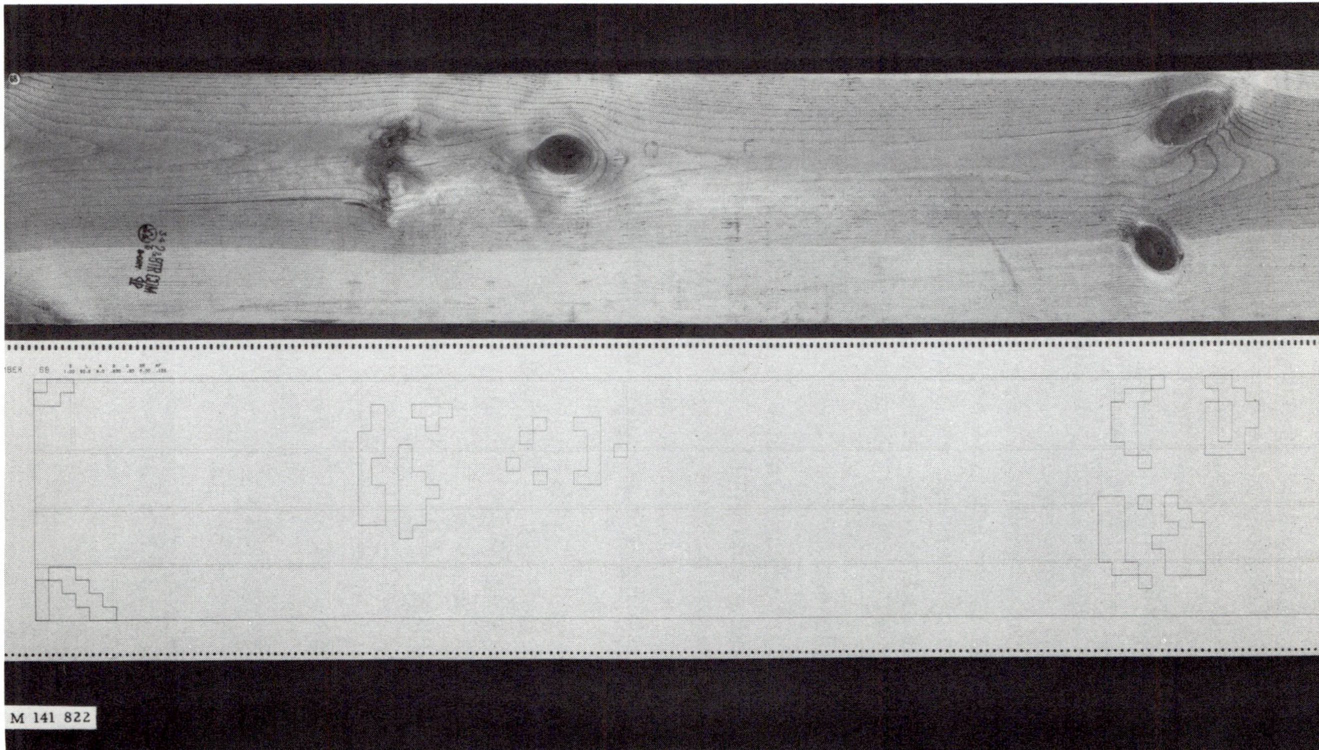

Figure 20.6 Computer plot of board also pinpoints location of clear cuttings through entire thickness of piece. Legend at top left of plot reads: $S = 1.00$; $L = 80.5$; $W = 9.0$; $G = 0.500$; $D = 0.50$; $SM = 6.00$; $KF = 0.125$.

required. This development is included in our work schedule.

While I am on the subject of schedule, it may be well to outline what remains to be done and when we expect to be finished. We are working with a single pair of transducers. This is not adequate for a production setup, and work with multiple transducers is underway. At present we are using water as a couplant between the transducer and the material being inspected. We are also investigating couplants other than water. Up to now, our work has concentrated primarily on sensing and locating knots.

Work is now underway to sense and locate the full range of defects. While previously we concentrated on an evaluation of transit time, the work we are now doing may be valuable in sensing defects such as decay or hidden pitch pockets.

Log readout planned

In our present research model of the system, we hold the piece being inspected and move the transducers. In a production model, it appears it would be better to hold the transducers and move the piece. Such changes will have to be worked out. With present funding and staffing, our schedule for completion of this work is about 1 1/2 years.

Then we hope to have specifications for a system that can sense and locate defects and a computer programmed to direct and control the sawing equipment to recover that yield. The system should operate at about 100 lineal feet per minute. Cost of the system is estimated at about $100,000. Payoff time on the investment should not exceed one year, depending on the application.

The system will be best suited for use in hardwood and softwood furniture cutup plants, hardwood and softwood millwork and molding plants, and at the edger and trim saws in both hardwood and softwood sawmills.

In the near future, we plan to begin work on a system to sense and locate defects in logs. This is a more difficult undertaking, and estimated completion date is about two years.

Will we see with sound to saw? Yes, I believe we will.

Exhibitors
1974 Sawmill & Plywood Clinic
Portland, Oregon

The many practical and innovative ideas offered by the 141 companies that exhibited at the 1974 Sawmill & Plywood Clinic added immeasurably to the Clinic's success. Registrants had the chance to speak with many exhibitors' representatives, receiving information focused on individual problems or simply gathering ideas for modifications under consideration. As registrants made their way from booth to booth, the opportunity to share ideas and reactions with other mill operators was extended beyond the Clinic auditorium.

Unfortunately, the specific information provided by the exhibits cannot be recorded here—but testimony from the many mill people viewing the show confirms its importance. The Clinic sponsors appreciate the support shown by the companies exhibiting in Portland. The following companies were represented at the 1974 Portland Clinic:

Action Equipment Co.
Agnew Environmental Products
Albany International Industries, Inc.
Applied Theory Assoc., Inc.
Archer Blower & Pipe Co., Inc.
Argo International Corp.
Armstrong Mfg. Co.
AutEx, Inc.
Auto-Chip Corp.
Automation Industries, Inc.

Bell Industries
Bigelow Machine Co., Inc.
A. B. Boyd Co.
Paul Bunyan, Jr. Enterprises

Cal-Ore Machinery Co.
Cascade Steel Fabricators, Inc.
C B Equipment Co.
Chip-N-Saw, Inc.
Coastal Machinery Co., Inc.
The Coe Mfg. Co.
Cothrell Machinery
Custom Equipment
Cutler-Hammer Inc.

Decision Dynamics, Inc.
Diamond Engineering Corp.
Disston, Inc.
Dodge Mfg. Co.
Durand Machine Co. Ltd.
Dynapar Corp.
Dyna-Tech Corp.

Eaton Corp., Industrial Truck Div.
The Ellison Co.
Elworthy & Co. Ltd.
Energex, Ltd.
ESCO Corp.

The Filer & Stowell Co.
Fluid Air Components Inc.
FMC Corp., Chain Div.
Furane Plastics, Inc.

Gardner-Denver Co.
Gardner Machines, Inc.
Gear Reducer Sales Co.
General Bearing Co.
Gillespie Decals, Inc.
Globe Machine Mfg. Co., Inc.
Graco Inc.
Griffith Polymers, Inc.

Walter E. Heller & Co.
R. Hoe & Co., Inc.
Howard-Cooper Corp.
Hydraulic & Air Equipment Co.

Industrial Electric Service Co.
Industrial Machinery Erectors Co.
Interlake Steel Corp.
Irvington-Moore

Jacksonville Blow Pipe Co.,
Div. of Montgomery
Industries Intl.

Knudtson Mfg. Co.

Letson & Burpee Ltd.
Lloyd Controls, Inc.
L-M Equipment Co., Inc.
Lumber Systems Inc. and
Progress Electronics Co.
of Oregon

MK-RDA, Inc.
Mac-Fab Mfg., Inc.
Mann-Russell Electronics, Inc.
Mark 50 Machinery, Inc.
H. C. Mason & Associates, Inc.
McCarthy Products Co.
McGrew Associates
McKay-Young, Inc.
Merco Products Inc.
Micro Switch Div., Honeywell Inc.
Microdyne Modular Electronics
Systems Ltd.
Mill Conversion Contractors, Inc.
Mitchell Bros. Industrial
Contractors, Inc.
Moore Oregon
A. W. Mortensen Co.

Morvue Electronics Systems
Munnell & Sherrill, Inc.

Newman Machine Co., Inc.
Nicholson Mfg. Co.
Noise Engineering Inc.
Nopak Inc.
North American Controls Inc.
Northwest Chain-Flow Systems
Northwest Dryer & Machinery Co.
Norton Co.

Opcon, Inc.

Pace Products, Inc.
Pacific Hoist
Pacific Saw & Knife Co.
Paulson Enterprises, Inc.
Ped Mfg. Ltd.
Peerless Div., Royal
Industries Inc.
Platt Electric Supply Inc.
Plymak Co., Inc.
Plywood Equipment Sales
Pneumatic Systems, Inc.
Portland Chain Mfg. Co.
Portland Iron Works
Portland Valve & Fitting Co.
Power Transmission Products, Inc.
Premier Gear & Machine Works
E. V. Prentice Co.
Production Machines, Inc.
Protective Coatings, Inc.

Rader Pneumatics Inc.
Ramic Products Div., Simco
Engineering Co.
Raute, Inc.
Raygo Wagner
Redco Controls, Inc.
Reid-Strutt Co. Inc.
Rema Electronic Ltd.
Rexnord Inc., Vibrating
Equipment Div.
Rogers Machinery Co., Inc.
Rolie's Machine Works, Inc.
Ronan Drying Systems, Inc.

Salem Equipment, Inc.
Signode Corp.
Simpson Timber Co.
Soderhamn Machine Mfg. Co.
Norman Springate &
Associates Ltd.
Star Machinery Co.
Stetson-Ross Machine Co., Inc.
Sullair Corp.
Superior Plywood Machinery, Inc.

Tallman Machinery Co.
Tech Serv Inc.
3M Co.
Top Line Equipment Co.
Triple/S Dynamics Inc.
Tyrone Hydraulics, Inc.

Ultrasonic Predictable
Maintenance Inc.
Unico, Inc.
United Tool Co.
Universal Plastics Co.

Wagner Electronic Products
Warner Electric Brake &
Clutch Co.
Wellons, Inc.
Western Machinery Corp.
Willamette Valley Co.

Index

Abrasive planing, 290-299
advantages of, 297
batch, 290-291
cost of, 296, 298
feed rates in, 292, 295
grade recovery with, 292, 296
in-line, 290, 291, 297
production with, 295-296, 298
Timesavers abrasive planer in, 290, 292
Adco West, 180
very small log mill (VSLM), 186
Administrative control of noise, 250, 253
Air flow noise, 267, 268-269
Alignment, of machine and machine components, 193, 194, 204-205, 206
Allocation of resources (for profit), 104
American Plywood Association, 97, 99
Amorphous powder, in Douglas fir bark, 91, 93, 95, 99, 100
as phenol substitute, 93, 99, 100
Automation. *See also* Scanners
of briquettors, 77-79
at debarker, 184
at headrig, 129, 178-180, 186
at quad saw, 128-129
in sorting, 129
of sticker stacker, 188

Baffles
acoustic, 258
in dry kiln, 246, 248
Bandsaw, 59
Bark
Douglas fir, 91-102
hogged, 66, 94
Bark burning, and slag problem, 234
Bast fibers, in Douglas fir bark, 91, 93, 95, 99, 100
for plastic manufacture, 93, 99
Batch system (abrasive planing), 290-291
Biles Coleman Lumber Co., 283
Bins, holding (storage), 180, 185
Board flow, 230
B.O.F.-handling (logs), 54
Bohemia, Inc., 91
Bohemia extender, 92, 96, 97, 98, 101
Boiler operation, 66
Breakdown techniques (logs to lumber), 53, 55, 59-60, 114. *See also* Sawing
Briquetting wood waste, 72-90
for conversion to electrical energy, 87-88
drying, 85, 86-87
in Europe, 74, 77-79
finances of, 78-79
firing, 86
history of, 73-74
presses for, 74-83
using coarse materials, 84-85
Bucking, 129
with computer, 111-113
with electric eyes, 224
of small logs, 184, 189
Bulkheads, for noise control, 252
Bureau of Land Management, 52
Burners. *See also* Energex Ltd.
fluidized-bed, 231-233
with pollution controls and blowers, 230

California Cedar Products, 205
Caran d'Ache Pencil Co., 79
Carriage feed device (shotgun), 168-170
Carriages, 59, 60, 149-167, 168-170
accuracy of, 150, 159-163
with computer, 156-159
cost of, 62, 164
history of, 150
precision, 60, 151-154, 159-167
with precision setworks, 154-167
reliability, 165
Cell wall soundingboard (wood), 310-311
Chip-N-Saw, 111, 118, 129
Chipping headrig, 108, 178-189. *See also* Chip-N-Saw; Headrig
Adco West VSLM, 186-189

automated (Chip-N-Saw), 129
with scanner and computer, 178-180
Chips, from lodgepole pine, 185
Circular-saw noise reduction, 267-273
Clark metering bin, 67
Closed-loop setworks, 154, 155-157, 160
Coleman points, 203
Computer, 103-118, 128-148, 149-167, 178-189. *See also* Minicomputers; Scanners
to control carriage, 60
to control setworks, 156-167
cost of, 118, 130, 145
as decision-making tool, 103-118
and linear programming, 104-118
maintenance of, 136-147
management with, 103-118, 134-135
and mill environment problems, 137-144
reliability of, 130, 135-136, 143
results, 118
and ultrasonic equipment, 311-315
Continuously rising temperature (CRT) drying, 300-307
Conversion. *See also* Wood, conversion
capabilities (machinery), 191-192
of logs to lumber, 54-55
of steam to water, 168
of waste to energy, 41-51
Conveyors
in briquetting installations, 83
fines, 70
hogged fuel, 67, 70
screw, 71, 83
Cork, in Douglas fir bark, 91, 93, 95, 98-99, 100
Cost
of abrasive planing, 296, 298
of computer, 118, 130, 145
of drying process, 301, 307
of lumber production, 62
of oriented fiberboard production, 124
of precision vs. conventional carriage/setworks, 164
of waste-fired power plant, 42, 44-49
of ultrasonic equipment, 318
Crown Zellerbach-Canada, 279, 280, 288
Culls, lumber from, 228
Cunit, 108
dollar value per, 111, 117
Cutterheads
on narrow lumber planer, 277, 279, 281-282, 284, 288
and noise reduction, 264-267, 279
Cutting program, 180, 185

Data collection, 235
Dadoing, 204-205
Dead timber, 189
Debarking, 180, 184-185
Decision Dynamics, Inc., 108, 118
Decision execution, by mill operators, 191, 195, 196
Decision-making
using computers, 103-107
instrumentation as aid for, 191, 195-196, 208
load on manager, 103
Defect observance, by ultrasonic equipment, 308-318
Defectoscope, in ultrasonic equipment, 311-312
Degrade, 235
kiln-induced, 302-303
reducing, 236-249
Diameter, 140
and recovery increase, 57-59, 108, 114, 146
of very small log, 178, 181
Die holder, in briquetting press, 81
Dimension lumber, 54, 55, 111
and abrasive planer, 297
breakdown patterns for, 114
green chain instructions for, 115
Dosimeters, 254
Downstairs flow, 230
Drum differential, 70
Drying. *See also* Kiln, dry
with abrasive planer, 295, 297
in briquetting process, 85-87
with continuously rising temperature (CRT), 300-307
cost of, for thick lumber, 236
of hogged fuel, 65-71
thermic, 85
Dust, as computer problem, 137, 140, 141-142, 147
Dyna-Tech Corp., 180
scanning system, 185

Edger
infeed to, 214-215
Portland Iron test, 209
quad band, 227-230
Efficiency, 190-210
and decision-making, 191, 195-196, 309
in mill operation, 58, 149, 190, 192-194
Electrical and electronic technicians, 175-177
Electrical process in wood fiber lab, 120-127
Electronics, 311. *See also* Computer
Employee utilization, 171-178. *See also* Maintenance; Operator
and design of control station, 172, 175, 178
for ideas to increase production, 173, 212-213
and reliability of control systems, 173
of technicians, 175-177
and working environment, 172
Enclosures (acoustical), 258-262, 267, 270-273
Energex Ltd., 70-71
Energy
conservation, with briquetting, 77-78, 87-90

Energy *(continued)*
recovery, with wood waste disposal, 41-51
Engineered wood products, 120
Engineering control
of drying in briquetting plant, 85-86
of noise, 250-252, 259-275
Environment, mill. *See also* Noise control
and computer problems, 137-144
and employees, 172
Eugene Water and Electric Board, 44
Evans Products Co. (Canadian Division), 108
Exposure meters to measure noise, 254

Fiber balance, 196
Fiberboard, 122, 124, 125
Fiber Orientation Process, 120-127
Fines, 70
Finger-jointed stock, and abrasive planing, 297, 298
Firing, in briquetting process, 86-87
Flash-drying, 87
Flow pattern in mill, 114, 184
downstairs, 230
smoothing, 211-235
upstairs (board), 230
Fly ash, 65, 66
Foam (acoustical), 256, 273, 274
Forklift
access to planer station, 218
importance of driver, 242-244
Forming station, in fiberboard plant, 124
Fuel. *See also* Hogged fuel
from solid waste, 44
from wood waste, 41-51
Furniture market, and ultrasonic defect detection, 315, 318

Gemaco water shotguns, 168
Glomera press. *See* Hausmann (Fred) press
Glue. *See* Bohemia extender
Goldberg, Rube (practices), 211-233
Grade, 190, *See also* Degrade
with abrasive planing, 291, 292
at green chain, 115-117
improvement of, 58-64
and trimming, 236
Green chain, sorting at, 115-116
Guides (for saws), 206

Hausmann (Fred) briquetting press, 72, 74, 76, 78-79
and breakdown of wood elasticity, 79-81
die holder of, 81
feeding device for, 79, 82-83
simplest, 82-83
Headrig, 59, 60, 129. *See also* Carriages; Chip-N-Saw; Chipping headrig
breakdown techniques at, 113-114
with water gun, 168-170
Hearing Conservation and Noise Abatement Committee of the Wood Products Section of the National Safety Council. *See* Racket Squad
Hearing conservation program, 250, 253
and proper hearing protection, 253
Histogram of accuracy, 192
Hogged fuel, 48, 49
drying system for, 65-71
with pollution-control burners, 231, 232, 233
salt content of, 67, 70
sand in, 71
Hula trim, 224
Human-engineered approach to production, 171-178

Installation technician, 177
Instrumentation, on-line, 208-210
Impact of cutterhead (noise)
magnitude of, 264-265
nature of, 265
velocity of, 264
In-line system (abrasive planing), 290, 291
Integrated circuit (I.C.) Chip, 176
Isolation, for noise reduction, 252, 258

Jack ladder, 186
John Deere hydraulic loader, 184
Joka Furniture Mfg. Co., 78-79

Kerf, 206
measuring, 196-197
minimizing, 191-192, 194, 196
and recovery increases, 53, 54, 57, 58, 59, 60
Kicking small logs, 185-186, 213
Kiln, dry, 188, 189, 239, 245-246
baffles in, 246, 248
and continuously rising temperature (CRT) technology, 300-307
cost of, 301, 307
maintenance of, 246
Kimwood Corp. abrasive planer, 297-298
Kink, 244-245
Knee face, 153
Knee-to-headblock mounting, 153-154
Knife molder, in abrasive planing, 298
Knot patterns, 55
location by ultrasonics, 313-315, 318

Lamb storage bin, 70
Length groupings
logs, 111
lumber, 237
LeTourneau log handler, 184
Light-emitting diode (L.E.D.), 176
Light, outside, as computer problem, 137
Linear programming, 104-118
Louisiana-Pacific Corp., 290, 298, 299
Lucite carriage, 198
Lumber carrier, 188, 189
Lumber recovery factor (LRF), 53
Lumber Systems, Inc., 283, 286-287
Lumber tally system, 129, 283

Mac-Fab Mfg., Inc., 180
Machinery
alignment, 193, 194, 204
allowance, 54
noise, 257
vibration, 193, 203
MacMillan Bloedel Ltd. (Canadian White Pine Division), 65
Maintenance, 150
of computer, 136-147
designing program for, 175
instruments for, 209-210
of kiln, 246
and mill environment, 137-144
training for, 143-144
Management, mill
complexity of, 105-107
with computers, 103-118, 134-135
purpose of, 104
Manufacturing improvements (lumber), 53-64
Market
for lumber, 52
for particleboard and fiberboard, 125
Measurement, 190-210
and efficiency of mill operation, 194-195
instrumentation for, 208
of kerf, 196-197
of operator performance, 195-196
in positioning, 199
of recovery, 111
of roughness, 197-198
of standard deviation of accuracy, 197
Mechanical extruder, 74
Merritt, Buck and Wayne, 214
Merritt Lumber Co., 214
Metering screw, 67, 70
Military-trained technicians, 176
Minicomputer, 165. *See also* Computer
with scanner, 111, 128, 131, 178, 185
MK-RDA, Inc., 119, 120, 127
Models (computerized) in mill management, 107, 108, 117-118
Moisture
in briquetting plant, 84-87
as computer problem, 137, 140
in fiber orientation, 121, 122, 125
in hogged fuel, 65-66
and ultrasonics, 311
Monitoring
cutting station performance, 208-209
lumber sizes, 226-227
saw performance, 204-205

Newsprint, briquetting of, 74
Noise control, 250-256, 257-275, 279
with enclosures, 258-262, 267, 270-273
by engineering, 251, 252, 257-275
with hearing protection, 253
by moving employees, 253
tooling for, 264-270
No-Nail Boxes, Ltd., 78
Northwest Hardwoods Co., 291
Northwest Noise Committee, 251

Observer technique, for measuring operator decisions, 195
Occupational Safety and Health Act (OSHA), 172, 250, 251, 253, 255, 257
Oil-bearing seed industry, 96, 101
Oil industry, 63-64, 72-73, 87-88, 95, 99-100, 101
Open circuit transistor collector, 176
Opening face, and measurement, 199
Open-loop setworks, 150, 154-155, 160, 164
Operator. *See also* Employee utilization.
and decision-making, 195-196
and efficiency, 194
relation to computer system, 111, 113, 146-147, 156-157, 159, 189
Optical coupler, 176
Optical laser systems, 204
Oregon State Board of Health, 251
Orientation of wood fibers, 119-127

Paint spotting, 227
Paper industry, 62, 64
Particleboard, 64, 122, 125
Patrick, Ken, 254, 255
Phenol substitute (amorphous powder), 93, 99, 100
Photographic observance of operator, 195-196
Piece count, 227, 236
Piling, 239, 248
Pineapple, 216
Planer-matchers, 276-277
for narrow lumber, 277-289
Planer roll, short stub, 218
Planing, 216-218
abrasive, 290-299
costs of, 248
with narrow lumber planer, 276-289
and noise reduction, 252, 273-275
and recovery, 53, 54, 57
and thick lumber, 236
Plywood, 117-118
adhesives for, 93, 97, 100
Polarization of wood fibers, 120-121
Pollution controls (air)
in bark breakdown, 101
using burners, 230-233
with hogged fuel, 65-71
with wood waste, 78
Portland Iron Test edger, 209
Positioning, measurement of, 199, 203
Power production, 41-51
from converted wood waste, 87
Power requirements
for briquetting, 83, 86
in closed-loop system, 156, 157, 165
for fiber orientation, 122
for mill operation, 194

Precision, in mill equipment, 59-60
ball screw, 60
carriage/dc setworks system, 149-167
sawing, 206, 290
test mandrel, 204-205
Presses. *See* Briquetting wood waste; Hausmann (Fred) briquetting press
Preventive maintenance. *See* Maintenance.
Process evaluation, 235
Production. *See also* Yield
with abrasive planing, 291, 295, 298
and employee ingenuity, 211-222
with small logs, 181
Product line, 53
Profilometer, 198
Profitability, 60, 178
computerized procedures for improving, 103-118
and efficiency, 58, 149, 190, 309
and log length, 111
Programming, linear
and mill management, 104-118
models from, 106-107
purpose of, 104-105
variables in, 106
Publishers Paper Co., 128-129, 131, 136

Quad band edger, 227-230, 234
Quad band headrig, 128-129, 136

Racket Squad, 251, 253
Recovery (lumber), 53-60, 108, 111, 178
with abrasive planing, 296, 298
of dead timber, 189
Regenerative solid-state power supply, 156, 157, 165
Regional power plant, 41-51
economics of, 42, 46-49
revenue from, 44
Resaw, 60, 227
Rockefeller, John Davison, 72-73
Roll cases, 218, 228
Roughness, 206
measuring, 197-198
minimizing, 191, 194, 196
Rubber, for noise reduction, 274

Salt, in hogged fuel, 67
Same-size output, 188-189
Sand
as base in fluidized-bed burner, 230-231
in hogged fuel, 71
Sawdust, drying of, 87
Sawing
live, 55-56, 114
for recovery increase, 54-62
variation, 53-54, 57
Saws
and collars for noise reduction, 273-274
computer-controlled, 111, 235
filing of, 193-194, 206
guides for, 206
noise reduction techniques for, 258, 267-274
tension, measurement of, 206, 209
vibration of, 193-194, 196, 203-205, 258, 267-268
Sawyer. *See* Employee utilization; Operator
Scanners, 128-148, 199, 235
reflective-type, 137
for very small logs, 178, 181, 185, 189, 213
Scragg, 213
Scribner log scale, 53
Sensor, on carriage, 151-153, 155-156
Sentry moisture detector, 280
Setworks, 60, 128, 149-167
accuracy of, 159-163
with computer, 156-159
functions of, 158-159
precision, 154-167, 194, 206
with precision carriage, 159-167
Sheet lead, for noise reduction, 273, 274
Shrinkage, and recovery, 53, 54, 57
Signode strapper and bander, 218
Silicon controlled rectifier (SCR) power supply, 156, 157, 165
Size. *See also* Same-size output
of dimension lumber, 54
monitoring, 226
Skate wheels, 214
Slabber, 228
Slag, 71
Slash, 178, 184
Slasher system, 184, 189
Slenderness ratio (of fibers), 122, 125
Small-log mill, 186. *See also* Very small log mill
narrow lumber planer for, 277-289
quad band edger as, 228
Sohn, Fred, 199
Solid waste, 41, 43-44, 49-50
Solvent, for wax extraction (mixed aliphatic aromatic solvent), 91-93, 95-96, 100, 101
Sorting, 115
automated, 129, 283, 286-287
to length, 237-239
of small logs, 178
Sound, and wood defects, 308-318
Southern Forest Products Association safety conference, 252
Sowards, Norman, 231
Spacers, 242, 244, 245
Species
birch, 305
cedar, 59, 108
western red, 65
fir, 53, 58, 108
balsam, 305
Douglas, 91-102, 228, 302, 305
white, 228
hemlock, 53, 58, 65, 93, 108
larch, 59
lauan, 52

maple, 305
oak, Mediterranean, 91, 95
pine, 52, 58
jack, 304, 305
lodgepole, 108, 185, 305
ponderosa, 305
southern yellow, 302, 305
redwood, 58
spruce, 59, 108
Canadian, 305
Engelmann, 305
Virola-like, 52
Spike knots, 55
Sprocket wheels, for lumber separation, 216
Stacking
with forklift, 242, 244, 248
with Lumber Systems stackers, 287-288
smoothing out flow in, 218-220
Standard deviation of accuracy, 206
measuring, 197, 210
minimizing, 191-192, 194, 196
versus range, 192
Standard Oil Co., 72-73
Statistical measurement, 195
Steam and power production, 42
Steam gun, 168, 170
Sterns Roger rotating dryer, 67
Stickers, alignment of, 239-242, 244, 245, 248
Sticker stacker, 186, 188
Stillman, Albert L., 79
Stop-and-loader, 186
Storage
of logs, 180, 185
of lumber, 225-226
Stumpage prices, 52
Sun Studs, Inc., 199
Surfacing (lumber), 54
Suspension dryer, 86-87
Sweed puller, 216

Temperature
and continuously rising temperature (CRT) technology, 300-307
dry-bulb, 305, 306
in hogged fuel drying process, 67-70, 87
of water gun, 170
wet-bulb, 305
Tension (saw), measuring, 206, 209
Thick lumber, disadvantages of, 236
3M Co., and abrasives, 297, 298
Timesavers abrasive planer, 290, 292
Tooth geometry, 206
Transducers, 312-313, 318
Transit time (ultrasonic), 312-315
Transport, 199, 201, 203
accuracy in, 193
Trimming, 222-224, 226, 236, 285
Troubleshooting, and equipment, 137-148, 176
Turning logs, 224-225
Twin resaws, 60, 227

Ultrasonics, 308-318
U.S. Department of Labor, advisory committee on noise, 250, 254
U.S. Forest Products Laboratory, 205
U.S. Forest Service, 52, 164
University of California, Department of Mechanical Engineering, 205
Upstairs flow, 230

Valon Kone debarker, 180, 184
Variables, in mill management, 106, 108
Vibration (saw), 193-194, 196, 203-205
as computer problem, 137
as noise problem, 258, 267-268
Very small log mill, 178-189
Volume recovery. *See* Yield

Walsh-Healy Act, 250, 251
Washington State University labs, 119, 120
Waste. *See also* Energy; Solid waste; Wood waste
disposal of, 41-51, 230
as fuel, 41-51
heat from, 78, 87
power from, 41-51
utilization of, 72-90, 178-189
Water gun, 168-170
Wax, in Douglas fir bark, 91-92, 94-97, 100
Weyerhaeuser Co., 131, 165-167, 197, 204, 205
and continuously rising temperature (CRT) drying, 302
Wood
conversion of, 62-64, 102, 119-127
as fuel, 64
as noise reducer, 256
recovery, 52-64
substitutes for, 308
total utilization of, 119-120
Wood waste
briquetting of, 72-90
disposal of, 41-51
as slash, 184
utilization of, 41-51, 290, 309
Workman's Compensation, and hearing loss, 254

Yield, 128, 129, 149-150, 162-164, 171, 190
definition of, 53
employees' role in increasing, 172-173, 178
improvement, 52-64
with ultrasonic defect observation, 315
York-Shipley Inc., 233

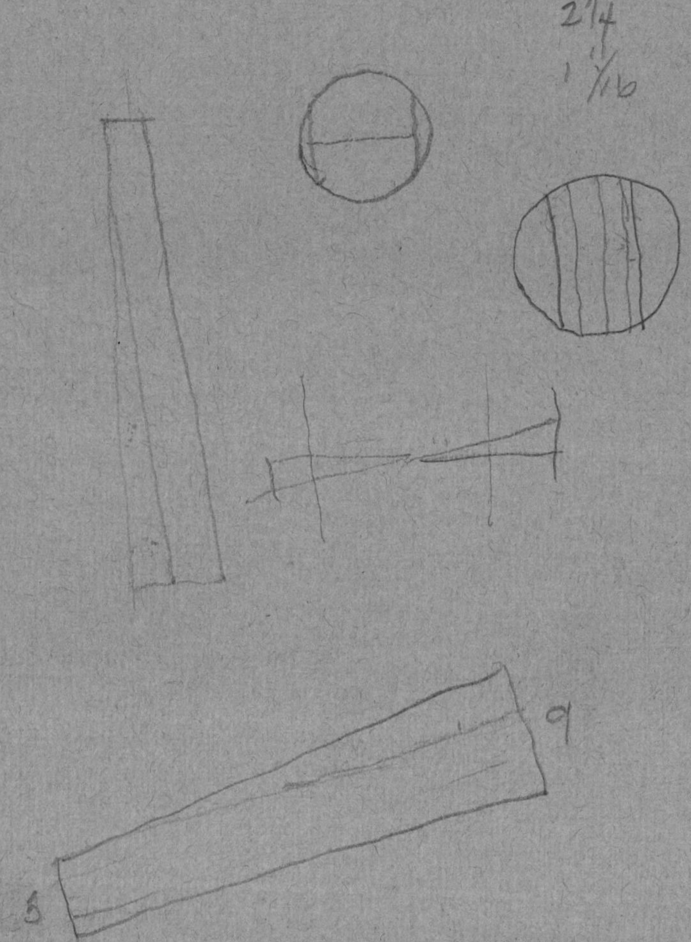
2 1/4
1 11/16
9
5